T. Lauder Brunton

Contributions to Our Knowledge of the Connexion

Between Chemical Constitution, Physiological Action, and Antagonism

T. Lauder Brunton

Contributions to Our Knowledge of the Connexion
Between Chemical Constitution, Physiological Action, and Antagonism

ISBN/EAN: 9783337248697

Printed in Europe, USA, Canada, Australia, Japan

Cover: Foto ©berggeist007 / pixelio.de

More available books at **www.hansebooks.com**

KNOWLEDGE OF THE CONNEXION

BETWEEN

CHEMICAL CONSTITUTION, PHYSIOLOGICAL ACTION,

AND ANTAGONISM.

BY

T. LAUDER BRUNTON, M.D., F.R.S.

AND

J. THEODORE CASH, M.D.

From the PHILOSOPHICAL TRANSACTIONS OF THE ROYAL SOCIETY — Part I. 1884.

LONDON:

PUBLISHED FOR THE ROYAL SOCIETY

BY TRÜBNER AND CO., LUDGATE HILL, E.C.

1884.

VIII. *Contributions to our knowledge of the connexion between Chemical Constitution, Physiological Action, and Antagonism.* [*]

By T. Lauder Brunton, *M.D., F.R.S., and* J. Theodore Cash, *M.D.*

Received June 13,—Read June 21, 1883.

[Plates 8–10.]

The great object of Pharmacology is to obtain such a knowledge of the relation between the chemical constitution and physiological action of bodies as to be able to predict with certainty what the action of any substance will be. One of the most important steps towards this object was made by Crum-Brown and Fraser, who showed that the introduction of methyl into the molecule of strychnia or thebaia changed the tetanising action of those poisons on the spinal cord into a paralyzing one on the ends of the motor nerves.

As the organic alkaloids are compound ammonias, it seemed probable that a similar change in the chemical constitution of ammonia itself might produce a corresponding change in physiological action. This was tested by Crum-Brown and Fraser, who found that trimethyl-ammonium iodide possessed a paralyzing action similar to that of methyl strychnia or methyl thebaia, while ammonia itself has been shown by Funke and Deahna to have a tetanising action very much like that of strychnia. A number of other ammonium compounds have been shown to have a similar paralyzing action; but there is no complete investigation of the whole series, nor has the relation of the acid with which the base is combined been determined.

In the present research we have attempted—

1st. To ascertain how the general action of ammonia is modified by its combination with an acid radical. Under this heading we have investigated: (a) the alteration in its general effects upon the organism; and (b) the alterations in muscle and nerve by which the general effects are to a great extent determined.

2nd. To investigate the general action of the compound ammonias containing the more common radicals of the alcohol series in the same way as the ammonium salts in the first part of the paper.

* The present research forms part of an investigation into the action of certain drugs on muscle and nerve, for which a grant was given to one of us (Brunton) in 1877, but the prosecution of which was much delayed by various circumstances, amongst others, the rebuilding of the laboratory in which the experiments were made.

3rd. To compare the action of ammonia on muscle and nerve with that of other substances nearly allied to it in chemical properties, and belonging to the group of alkalies.

4th. To examine the action of acid and alkali upon muscle independently of the chemical composition of the acids or alkalies employed.

5th. To extend the research on muscle and nerve to the elements belonging to the group of alkaline earths.

GENERAL ACTION OF AMMONIUM SALTS.

From experiments with ammonium chloride, sulphate, phosphate, tartrate, benzoate, and hippurate, FELTZ and RITTER concluded that ammoniacal salts all had a similar action, producing convulsions and coma, slowing of the pulse and lowering of the temperature. They considered the action to be the same in kind, but differing in intensity. The convulsions produced by ammoniacal salts were shown by FUNKE and DEAHNA to be similar to the tetanus produced by strychnia, differing from it only in the fact that a single convulsion instead of a series of convulsions was produced by the poison. The cause of this result they believed to be the rapid production of paralysis of the motor nerves by the ammoniacal salt, which prevented the occurrence of more than one tetanic convulsion.

As the action of chloride of ammonium has already been pretty thoroughly investigated, it seemed to us unnecessary to make any more experiments upon its general action. We have therefore restricted our researches to the action of the bromide, iodide, sulphate and phosphate, and have experimented only on Frogs with the bromide. The result of these experiments seems to be that ammonium chloride, bromide and iodide form a series. At one end of it is ammonium chloride having a stimulant action on the spinal cord, and, at the other, the iodide having a paralyzing action upon motor nerves. Ammonia and ammonium chloride produce tetanus; the bromide, hyperaesthesia, with some clonic spasm, passing into tetanus, which, however, comes on very late in the course of the poisoning. The iodide produces rapid failure of higher reflexes, such as that from the conjunctiva, and caused in our experiments progressive paralysis, but no tetanus. At an early stage of poisoning by it the Frog responded with a creak when stroked on the back, and as this has been shown by GOLTZ to occur after removal of the cerebral hemispheres, its occurrence in poisoning by ammonium iodide may be looked upon as a proof that the higher centres are poisoned first. After injection of ammonium phosphate also, there is throughout an absence of true spasm. The usual movements become sprawling, and when taken up and gently set down again, the animal remains plastic, with the limbs extended. Before the cessation of reflex in the hind limbs, slight twitchings are observed to accompany induced movement. After the injection of sulphate of ammonium a slight degree of hyperaesthesia is developed. In a variable length of time twitchings occur. They appear first in the anterior extremities, and then spread all over the body to the hind limbs. This spasm increases in intensity, and often manifests itself by a number of clonic convulsions occurring at tolerably regular intervals. These seldom pass into a rigid tetanus. They are, however, provoked by touching the animal, by the application of cold to the surface of its body, or by a blow upon the table upon which it is resting. When the sciatic nerve was divided on one side before the injection of the poison, twitchings did not occur upon that side. The action of the salts of ammonia upon the circulation was also found to be various. Thus, in poisoning by the bromide, it was unusual to find the heart materially influenced in its activity, even when the most marked motor symptoms had been developed. With the iodide, however, an early arrest of the heart in diastole, with the auricles and ventricle distended by dark blood, was very usual. A larger dose of the phosphate, and not unfrequently an equal dose of the sulphate, had a somewhat similar effect. An examination of the blood showed that after poisoning by bromide of ammonium, a marked change had taken place in the red blood-corpuscles. These exhibited numerous coagulations in their stroma; an increase of free nuclei was likewise observed in the blood; where the blood from the corresponding limb to which the poison had not had access was examined, no such changes were observed. A similar result is occasionally noticed after poisoning by the sulphate; it is much more unusual where the iodide and phosphate have been employed.

Examination of the reaction of the muscle to direct and indirect stimulation was made as rapidly as possible, when it was desired to examine their reaction at any stage which the poisoning had reached. The ligatured limb was used for a contrast; and as it has been shown by KÜHNE[*] that in cold-blooded animals the irritability of the muscle declines when containing blood in a condition of stasis, allowance must be made for this decrease in irritability when contrasting its reaction with that of the poisoned muscle. The irritability was tested by means of approximating the secondary coil of a DU BOIS REYMOND'S inductorium to the primary, the greatest distance at which a minimal contraction was produced being registered both for direct and indirect stimulation. This figure was controlled by removing the secondary coil from the primary, in which case contraction often persisted at a more distant position than it was observed at when the coil was approximated.[†] The muscle poisoned by bromide showed an increase of irritability in the early stages, and before the action of the poison was complete. There was a slight but less marked increase occasionally in the case of iodide, but usually the irritability in cases of slight poisoning is diminished. There is usually no marked increase of irritability in muscles poisoned by the phosphate and sulphate, though in exceptional cases it has been observed as a temporary condition in both The muscle responds to direct and indirect stimulation (opening shock) by a long, at first equally high, but then rapidly falling curve, in comparison with the normal. The

* Archiv. f. Anat. u. Physiol., 1859.
† The excitability of the muscle appearing to be increased by its contraction.

response to indirect stimulation is, however, much feebler than to direct. The tetanus of both is impaired, but especially that of indirect stimulation. The total failure of reaction upon stimulation of the nerve frequently occurs whilst the muscle yields a moderate tetanus. If the heart has not been arrested by the injection of too large a dose of ammonium iodide before the circulation has distributed the poison sufficiently, it is often found that stimulation of the nerve does not produce any contraction, or it may be only a few faint twitches of the muscle. In poisoning by the phosphate of ammonium direct stimulation produces, as a rule, a tolerably good, though prolonged contraction, but the failure of reaction to direct and indirect stimulation is more parallel than in poisoning by the iodide, and if the irritability of the nerve is entirely lost, it is usually found that the muscle when stimulated directly contracts but very feebly even to the strongest tetanising current. Ammonium sulphate paralyses both muscle and nerve. The reactions given by the former are, however, longer, and outlast those of the latter. The tetanus curve of both is feeble, even in cases of rapid poisoning.

ACTION OF COMPOUND AMMONIAS.

Our experiments with these bodies were made upon frogs, rats, and rabbits. The substances employed, twenty-six in number, were :—Ethylamine, trimethylamine, triethylamine ; the chlorides of methyl-ammonium, ethyl-ammonium, amyl-ammonium, dimethyl-ammonium, diethyl-ammonium, trimethyl-ammonium, and triethyl-ammonium ; the iodides of methyl-ammonium, ethyl-ammonium, amyl-ammonium, dimethyl-ammonium, diethyl-ammonium, trimethyl-ammonium, triethyl-ammonium, tetramethyl-ammonium, and tetraethyl-ammonium ; the sulphates of methyl-ammonium, ethyl-ammonium, amyl-ammonium, dimethyl-ammonium, diethyl-ammonium, trimethyl-ammonium, and triethyl-ammonium. The action of all these bodies was tested in Frogs, but the whole of the series was not investigated in Rats and Rabbits. All the salts of the compound ammonias which we used were obtained from Messrs. Hopkins and Williams, who prepared them expressly for us, and guaranteed their purity. The poison was in all cases administered by subcutaneous injection.

We have compared first the action of the compound ammonias, uncombined with an acid radical, with the action of ammonia itself. We have then compared the actions of the chlorides, iodides, and sulphates, of the compound ammonias with each other, and with the corresponding salts of ammonium. It will be noticed that there is a considerable difference between the action of the compound ammonias and of ammonia. The tendency to produce tetanus resembling that of ammonia was noticed in ethylamine, which was the only one of the compound ammonias containing only one atom of hydrogen, replaced by a radical, that we investigated in a free state, uncombined with acid. When used as a chloride, the convulsive action was less marked. The substitution of even a single atom of hydrogen by an alcohol radical appears to lessen the tetanising action of ammonia, and this diminution is increased by the substitution of two or three atoms, then a change takes place, and when the ammonia is combined with four atoms of an alcohol radical, a convulsant action again becomes more marked, though it is not so great as in the case of ammonia itself.

With these exceptions, the symptoms were those of gradual motor paralysis. This motor paralysis appeared to us to be due, in a great measure, to a paralyzing action of the substance on the spinal cord, as motion ceased in the animal at a time when the muscles and motor nerves were still capable of vigorous action.

The tetramethyl- and tetraethyl-ammonias appear to have a particular tendency to paralyse the higher reflexes before the lower, so that reflex from the cornea disappears sooner than from the foot. They appear also to affect the heart more than the other compound ammonias, so that in poisoning by them the heart was generally found motionless, in complete diastole, and distended with dark blood.

We did not observe the same marked difference between the action of the different salts of the compound ammonias that we did in the case of ammonia itself. The iodides, however, appear to affect the heart more powerfully than other salts, and to cause its arrest in diastole.

The chlorides and sulphates also appear to have a greater tendency to produce muscular tremor than other salts.

We have drawn up, in a tabular form, an epitome of the symptoms of poisoning produced by salts of the compound ammonias in Frogs, Rabbits, and Rats. The tables may appear bulky, but the number of salts experimented upon was great, and as they were difficult to prepare, and expensive to procure, we have thought it advisable to give an example of the general action of each drug, as well as a summary of the results which we have obtained. We have, however, put them as shortly as possible, and restricted ourselves to one experiment with each substance on each kind of animal.

ACTION of Simple and Compound Ammonias on Frogs.

Substance	Dose.	Symptoms.	Post-Mortem Appearances and Reactions.
(a) Ammonium bromide	0·25 to 0·5 grm.	In some cases hyperæsthesia, followed by spasmodic twitchings, succeeded by well marked tetanic spasm. The last symptom is late in appearing, 15' to 25'. Heart usually active throughout.	Heart is usually beating. There are many convulsions in red corpuscles. In early stage of poisoning, there is marked irritability of poisoned as compared with non-poisoned muscle. Later, the M. and N. curve of the poisoned limb attain length, their response in tetanic contract more feeble. Eventually this condition increases to total failure of reaction of the nerve. This occurs when the muscle still reacts to direct stimulation, but its performance is weaker, especially its tetanic curve, and a stronger shock is needed to excite its activity than that needed for the normal muscle. The paralysis is not due to exhaustion from tetanus, as it occurs when tetanus has not been a prominent symptom. The change of corpuscles and the continued action of the heart are constant.
(b) Ammonium iodide	0·25 to 0·5 grm.	Higher reflexes quickly disappear. Limb reflexes persist longer, and require stronger stimulation to elicit them. There is increasing sluggishness, but no spasm; in fact, a striking absence of nervous symptoms.	Heart is usually at rest in diastole, containing dark blood. It is unusual to see convulsions in red corpuscles. It seems doubtful whether muscular irritability is increased as a temporary condition. If the heart has not been arrested too soon by the extent of the dose, the nerve soon becomes deeply affected by the poison, and only yields two or three faint response to the strongest stimulation. Muscle-curve, though prolonged, is not, as a rule, extensively impaired. Frequently the muscle contracts well when irritated directly, either by a single shock or tetanic irritation, when the nerve refuses to react to the strongest stimulation.
(c) Ammonium phosphate	0·1 to 0·75 grm.	Increasing sluggishness, with gradual failure of reflex. Squeaking movements on stimulation, but the animal tends to remain placid. There is an entire absence of active nervous symptoms, except occasionally a slight twitching in drawing up leg before reflex manifestation has ceased.	In any but cases of very slow poisoning, heart is found in diastole arrest and full of dark blood. The red corpuscles do not usually show convulsions. There may be a slight and transitory increase of irritability in early stage of poisoning. Nerve and muscle both tend to fail on poisoned side, and that pretty equally. In two cases the nerve did not respond at all, and the muscle only by very feeble curve to strongest tetanus. Heart still contracted feebly in answer to stimulation.
(d) Ammonium sulphate	0·25 to 0·5 grm.	Movements soon accompanied by twitchings, and in 30° to 40° clonic spasm becomes established. A lasting tetanus is decidedly rare, but yet occurs. Clonic spasms commences in arms and chest. There may be 5 to 7 of these short spasms in the minute.	Heart frequently in diastole stillstand, with very dark blood. Red corpuscles of blood only very occasionally show slight convulsions. The muscle and nerve appear both to be paralysed, the nerve, however, giving way first. The tetanus curve is feeble in both, even in cases of rapid poisoning. If the sciatic be cut so as to prevent exhaustion from spasm, the reaction of muscles from the two limbs is equal.

* M. stands for direct stimulation of the muscle itself, N. for indirect stimulation of the muscle through its nerve.

ACTION of Simple and Compound Ammonias on Frogs (continued).

Substance.	Dose.	Symptoms.	Post-Mortem Appearances and Reactions.
1. Ethylamine	About ·05 grm.	Circulation becomes slow. Reflex slow apparently, then spasm chiefly in pectoralis and abdominal muscles is noticed, and this develops into a well marked and general tetanus.	Heart beating. Irritability of N.* and M. increased as compared with ligatured limb. Thus strong tetanus of poisoned at 6 centims. induction coil, while that of ligatured limb was only partially developed. Contracture tends to increase rapidly on repeated stimulation. In another case curve of N. and M. prolonged. N. soon gives way in tetanus.
2. Trimethylamine	About ·05 grm.	Gradual failure of circulation and of reflex. Pigment cells dilated. No spasm of a tetanic character, though some clonic contractions towards the end of the poisoning.	Heart in diastolic still-stand, and containing dark blood. Both M. and N. yield strong contraction and tetanus. The tetanus of the N., however, seem tends to cloose.
3. Triethylamine	About ·10 grm.	Slowing of circulation; dilatation of pigment, cells of web. Gradual failure of reflex, unaccompanied by spasm.	Heart beating feebly, with tendency to diastolic arrest. Blood smells strongly of triethylamine. Both M. and N. cause contraction freely. Poisoned muscle and nerve are rather more irritable than the ligatured. Tetanus is more extensive at first, but soon tends to give way. There is increased viscosity from M. on repeatedly stimulating by induction shocks.
4. Methyl-ammonium chloride	·2 ·3 grm. of a saturated solution	Gradual failure of circulation and of reflex, generally unaccompanied by spasm. Jerks of lower limbs is longer preserved.	Heart tending to diastole, or in diastolic arrest, and containing dark blood. M. yields a slightly prolonged curve. Nerve, as a rule, reacts but faintly to strongest stimulation. The tetanus of N. is imperfect and very feeble.
5. Ethyl-ammonium chloride	About ·03 grm.	Reflex becomes gradual, slow, and spasmodic, though there is no genuine tetanus. Just before its abolition pinching of toe causes a general though faint reflex of all body, without withdrawal of foot. Circulation continues moderately good.	Heart beating. Muscle yields a good curve, which alters, however, rapidly increasing contracture on repeatedly stimulating. The tetanic (M.) curve is also good, though followed by very slow relaxation. On strong stimulation there was no N., though strong tetanising current caused a very feeble and transient contraction.
6. Amyl-ammonium chloride	3 grt. sal. sol.	Excitement at first, with acceleration of heart. Lower reflexes persist. There is no spasm. Higher reflexes soon lost.	Heart beating in vermicular manner, does not empty itself. Curve from M. tends to shorten rapidly on repeated stimulation. Tetanus curve strong, though not equal to normal. N. yields only a few contractions to strongest stimulation. Its tetanus is feeble, and rapidly diminishes.
7. Dimethyl-ammonium chloride	·? grm.	Becomes tremulous and weak. Gradual failure of reflex.	Heart beating. Curve of M. and N. lengthened and higher, and tends to become elongated more rapidly than normal.
8. Diethyl-ammonium chloride	·05 ? grm.	Increasing lethargy with failure of reflex. At last slight tremor of foot and of all body occurs on irritating foot.	Heart in diastolic still-stand. Beats with vermicular movement if irritated. Blood dark. Red corpuscles show considerable coagulation, whilst some are to be seen in ligatured limb. Curve of both M. and N. prolonged, and especially N. soon fails. Tetanus of N. soon collapses.

* M. stands for direct stimulation of the muscle itself, N. for indirect stimulation of the nerve through its nerve.

Action of Simple and Compound Ammonias on Frogs (continued).

Substance	Dose.	Symptoms.	Post-Mortem Appearances and Reactions.
9. Trimethyl-ammonium chloride	.25 grm.	Circulation good. No spasm. Increasing lethargy.	Heart beating. Curve of both M.* and N. more extensive and longer than normal.
10. Triethyl ammonium chloride	.125 grm.	Circulation continues good. No spasm. Attempt to spring, but no force in the movement. Gradual loss of power and reflex manifestations.	Heart beating. No abnormality. Muscle and nerve more irritable to minimal stimulation. The curve, though strong at first, rapidly falls on repetition, and length no considerably.
11. Methyl-ammonium iodide	.15 grm.	Increasing torpor. Heart slowed.	Heart beating slowly. N. and M. more irritable. Give ordinary curve.
12. Ethyl-ammonium iodide	.15 grm.	Increasing sluggishness. Croaking reflex. Respiration and heart good.	Heart beating. N. and M. more irritable. Give ordinary curve. In another case total paralysis of N. with prolonged and bumped curve.
13. Amyl-ammonium iodide	.15 to .2 grm.	Circulation tends to slow. Leg drawn up with jerking (staccato). Becomes difficult to excite reflex.	Heart beating slowly. Some twitching of legs still. Muscle and nerve irritability is increased in relation to the other limb. N. gives a feeble contraction, but with double hump. Muscle gives an extensive contraction, much longer than normal, and with distinct double hump. It is of normal altitude.
14. Dimethyl-ammonium iodide	.1 grm.	Increasing apathy. No spasm.	N. soon fails in tetanus. Both M. and N. curve prolonged.
15. Diethyl-ammonium iodide	.2 grm.	Circulation good. Croaking reflex. Tends to lethargy. No spasm. Colour dark.	Heart beating slowly. Minimal irritability of N. and M. distinctly increased. Single curve of each like normal, but on stimulation being repeated a distinct double hump occurs in both cases, with rapid lengthening of the curve.
16. Trimethyl-ammonium iodide	.15 grm.	Becomes gradually sluggish. Movements more short, feeble, and staccato.	Heart full of dark blood—still acted. Red corpuscles show coagulation to some extent, the blood of ligatured leg being normal. Nerve gives curve which rapidly falls to insignificant proportions, becoming humped. M. gives a strong contraction, with a most distinct hump. On repeated stimulation the muscle shortens rapidly.
17. Triethyl-ammonium iodide	.1 grm.	Increasing sluggishness and gradual loss of reflex.	Heart beating slowly, with vermicular movement. Red corpuscles show slight coagulations. Irritability of M. and N. diminished. Both yield a longer lower curve, rapidly tending to fail. Tetanus of N. is very weak.
18. Tetramethyl-ammonium iodide	.007 to .028 grm.	Spasmodic twitchings of trunk and leg muscles. Limbs drawn up in very tremulous manner. Whole body twitches in response to pinching foot, even when the foot is no longer withdrawn. Death within a minute with finger done.	When all reflex has ceased but faint tremor, the N. may still respond well.

* M. stands for direct stimulation of the muscle itself, N. for indirect stimulation of the muscle through its nerve.

Action of Simple and Compound Ammonias on Frogs (continued).

Substance.	Dose.	Symptoms.	Post-Mortem Appearances and Reactions.
19. Tetra-ethyl-ammonium iodide	.007 to .072 grm.	As in tetra-methyl-ammonium iodide.	Heart in diastole still-stand, engorged with dark blood. The M.* yields usually a good curve to powerful stimulation. This tends to lengthen rapidly on repeating stimulation. In one case the muscle was almost completely paralysed. The nerve, when heart has not been too soon arrested, is profoundly paralysed.
20. Methyl-ammonium sulphate	.3 grm.	Slowing of circulation. Increasing lethargy. Reflex falls to faint tremble without withdrawal of leg.	Heart beating. Red corpuscles almost all disappeared. Nuclei free. N. gives no response to strongest tetanus. M. still irritable, gives a low prolonged curve. It gives a tetanus, at first firm, but soon collapsing.
21. Ethyl-ammonium sulphate	.2 grm.	Slowing of circulation. Increasing sluggishness. Slow answer to stimulation by reflex movement. Reflex gradually reduced to tremor.	Heart beating slowly (12 per min.). Both systole and diastole slow. Minimal irritability of M. and N. greater. Curve tends to lengthen rapidly. Both give strong continuous tracing.
22. Amyl-ammonium sulphate	.1 grm.	Slowing of circulation. Increasing sluggishness. Leg drawn up tremblingly and incompletely. Leg reflex ceases.	Heart beating 24, feebly. Systole long, and systole slow and still. Minimal irritability of nerve much diminished. N. yields good curve to powerful stimulation, but it tends to fall rapidly on repeating stimulation. Muscle gives a strong curve, with rapidly increasing break. Continuous curve of M. is much shorter than that of normal muscle.
23. Dimethyl-ammonium sulphate	.25 grm.	Heart's activity impaired. Tremulousness and weakness increase. All reflex eventually ceases in leg.	Heart beating slowly. Coagulation in all red corpuscles, and some free nuclei. No response of N. to strongest tetanus. Muscular irritability impaired. Curve is long and low, and its tetanus is very feeble, whilst that of ligatured limb is very extensive.
24. Diethyl-ammonium sulphate	.25 grm.	Increasing torpor, but if sufficiently excited gives strong extension with long latency. Slightly tremulous.	Heart beating slowly. Red corpuscles full of coagulations. Poisoned N. and M. more irritable. On continuous tetanus N. tends to pass into clonus.
25. Trimethyl-ammonium sulphate	.25 grm.	Tremulousness develops. All movements become rapid, tremulous, and uncertain. Leg at last drawn up by a sharp and unsustained contraction, or by a series of twitches.	Heart beating well. Red corpuscles all show coagulations, many free nuclei. Poisoned leg gives good curve, though minimal irritability is much impaired. In tetanus, however, nerve soon gives way and passes into clonus.
26. Triethyl-ammonium sulphate	.2 grm.	Irritability appears at first to be increased. Spring becomes tremulous, staccato and weaker. Spasmodic movements continue some time after stimulation. At last stimulation of foot causes slight trembling of foot and whole body, but no withdrawal of foot.	Heart beating slowly, much engorged, scarcely empties itself at all. Red cells all contain coagulations. There is no change in minimal irritability, and none in curve, except that there is tendency for altitude to fall in case of poisoned. N. tetanus is also weaker. In another case total paralysis of nerve.

* M. stands for direct stimulation of the muscle itself, N. for indirect stimulation of the muscle through its nerve.

Action of Compound Ammonium on Rats.

Solution injected.	Quantity gradually injected into Abdominal Cavity.	Symptoms in Brief.	Ultimate Result.	Post-Mortem Appearances and Reactions.
1. Diethyl-ammonium chloride	·5 grm.	Walk becomes straddling and ataxic, movements become jerking and tremulous. Attacks of shaking provoked by movement. Pey with lower jaw on table, and extensor-jerk back head in a clonic manner. Dyspnœa, laboured breathing. Runs backwards. Facial muscles spasmodically contracted.	Death.	Heart beating. Right side of heart full.
2. Triethyl-ammonium chloride	·6 grm.	Listless. Occasionally raises itself in a spasmodic manner. Slight tremulousness. Rapid flexure of head, causing rapping on table. Tremor becomes general. Springs upwards and backwards. Spasm more violent. Some dyspnœa. No symptom of pain.	Death.	Heart beating. Right side of heart full. Left empty.
3. Methyl-ammonium iodide	·2 grm.	Quiet. Waddles in walk. Movements sprawling. Somewhat anæsthetic. Progress slow and wavering. Hind limbs much paralysed. Tail sometimes individually thrown up. Occasional twitching of tail and limbs, but no general convulsion. General body reflex from stimulation of hind limbs, but hardly in alarm.	Death.	Heart contains some blood. Mesenteric vessels somewhat dilated. Stimulation of nerve gives hardly any contraction, but direct stimulation gives fine contraction.
4. Ethyl-ammonium iodide	·5 grm.	Apathetic, increasing weakness. Power soon begins to fail in hind limbs, tends to fall over on side when walking. Respiration slow. No spasm. Withdrawal of fore feet if hind touched, but not of latter. Reflex only in fore part of body.	Death.	Heart and lungs normal. Mesenteric vessels not dilated.
5. Amyl-ammonium iodide	·	Twitchings of hind legs, with occasional active extension. Head rapidly flexed and extended, so that jaw raps on table. Movements sprawling. Dyspnœa. Rests on belly. Falls on side. Spasm of legs and body, and death. There was no true tetanus.	Death.	Heart contains little blood in left side. Right side somewhat congested. Dyer early and strong. No mesenteric inflammation.
6. Dimethyl-ammonium iodide	·5 grm.	Torpidity. Sways after movement. Falls over on side repeatedly when walking. No spasm. No anæsthesia. Gradual failure and death.	Death.	Right side of heart full. Left empty. No mesenteric inflammation.

ACTION of Compound Ammonias on Rats (continued).

Solution Injected.	Quantity gradually injected into Abdominal Cavity.	Symptoms in Brief.	Ultimate Result.	Post-Mortem Appearances and Reaction.
7. Diethyl-ammonium iodide	.4 grm.	Gait becomes feeble and wavering. Sways when trying to sit up. Scrambles with feet in order to preserve balance. Accasional teetantaneous twitch in back and fore limbs observed, which gives the impression of a hiccough. Lies prone on belly. Spasmodic twitchings, 8 or 10 per minute, occur.	Killed.	Heart normal. There is no great change in excitability of muscle to direct or indirect stimulation.
8. Trimethyl-ammonium iodide	.2 grm.	At once powerful convulsions (no tetanus). Dyspnea. Falls on side and dies (action very rapid). A much smaller dose fatal in other cases.	Death.	Brain not markedly congested. Abdominal vessels dilated.
9. Ethyl-ammonium sulphate	.5 grm.	Torpid. An occasional heaving of body (like hiccough) and throwing up of head. Lachrymation. Movements gradually become tremulous. Rocks when walking. Breathing accelerated. No spasm. Ceases to move at all, except when disturbed.	Killed.	Right heart full. Left empty.
10. Amyl-ammonium sulphate	.15 grm.	Walking slightly tremulous. Soon violent trembling of head and fore part of body, increased on movement. Dyspnea. Lies on belly. Anæsthesia. Scrambling with feet to maintain equilibrium. Runs forward rapidly, and slews short. Gait like paralysis agitans. Powerful general convulsion, extension of legs, and jumping from side to side. Runs backwards. Falls on side.	Death.	Heart beating 60 per 1'. Both muscle and nerve respond well and equally to induction shock. Some congestion of brain, of membranes of cord, and of cord itself.
11. Diethyl-ammonium sulphate	.6 grm.	Movements straggling, tremulous, like paralysis agitans. Rocks from side to side. Flexion and extension of head; "rapping" develops and become frequent, often accompanied by slewing of all body. Runs with difficulty if laid on side. Rapping ceases. Respiration becomes very feeble. Reflex gradually lost.	Death.	Muscle and nerve are both irritable. Heart normal. No peritonitis.

Action of Simple and Compound Ammonias on Rabbits.

Solution injected.	Quantity.	Symptoms, with Notes of Time of their Occurrence in Brief.	Ultimate Result.	Post-Mortem Appearances.
1. Trimethyl-ammonium chloride .	1 grm.	In 35ᵐ lies on belly. Breathing slow. 1ʰ is approaching normal, but is weak, and rocks if moved. Found dead in morning.	Died.	Kidneys are much congested. There is a good deal of yellowish viscid fluid in the intestines. Bladder is full of a greenish, deeply-coloured urine, containing much white flocculent matter. Brain and cord are congested. Liver is much congested. Death may perhaps be attributed primarily to the ailing condition of the animal. The symptoms due to the poison were very slight, the chief one being increasing weakness.
2. Diethyl-ammonium chloride .	1 grm.	In 25ᵐ movements of legs rather staccato, but soon assume normal character.	Recovered.	
3. Trimethyl-ammonium chloride	1 grm.	In 25ᵐ lies with fore legs extended. Head rocks if touched. Much salivation, which continues all through. Pupil contracted. 1ʰ head tends to fall on table. 2ʰ 5ᵐ lies with fore legs fully extended on either side. 3ʰ 35ᵐ lies on side with legs in all directions. Tries to escape if approached. No anæsthesia. Better almost lost in hind limbs. 5ʰ 35ᵐ rather in all limbs almost entirely lost. There has been no tree spasm in this case.	Died.	There is no peritonitis. No unusual amount of fluid in peritoneal cavity. Right heart full; left empty. Lungs and kidneys normal. Liver congested. Brain and cord not congested.
4. Triethyl-ammonium chloride .	1 grm.	In 1ʰ restless. Scrambling with fore feet. Movements staccato. Trembles if touched. 2ʰ 5ᵐ a violent attack of scrambling movements. If allowed to run, movements are tremulous and associated with much involuntary movement, 4ʰ head down, rump up. No anæsthesia. Drumming movements with fore limbs. Convulsive movements appear clonⁿ᷉er movement is attempted. The strength of the animal never seems much impaired. 7ʰ decided improvement in condition.	Recovered.	
5. Triethyl-ammonium iodide . .	2 grms.	1ʰ 3ᵐ hastight tremor. 1ʰ 10ᵐ tremor very distinct on provoking movement. Hind legs are thrown out in hopping in a tremulous manner. 2ʰ 13ᵐ there is now no tremor or weakness when animal is taken up.	Recovered.	
6. Ethyl-ammonium iodide . .	2 grms.	The second injection of 1 grm. was 2ʰ 10ᵐ after the first. After the first injection 22ᵐ drowsy. Sits in normal position, with an occasional slight tremor. 60ᵐ trembles on being taken up and then settles down again. 1ʰ 55ᵐ tremor has ceased. Injected 1 grm. Increasing apathy, ceases to notice other rabbits. 1ʰ32ᵐ after second injection appears somewhat anæsthetic. 2ʰ 5ᵐ remains sitting perfectly still if placed on floor. Shakes if compelled to move.	Died.	Brain and its membranes not markedly congested. Right heart full; left empty. Intestines full of a brown fluid, with fæcal odour. No peritonitis.

ACTION of Simple and Compound Ammonias on Rabbits (continued).

Solution Injected.	Quantity.	Symptoms, with Notes of Time of their Occurrence in Brief.	Ultimate Result.	Post-Mortem Appearances.
7. Amyl-ammonium iodide	1 grm.	In 15ᵐ there is distinct trembling of all body. 1ʰ tends to sink upon belly, but soon recovers itself. Hind legs specially affected. 1ʰ 30ᵐ lies several seconds with legs fully extended. In 2ʰ 20ᵐ drumming and scrambling movements. Cannot remain more than 2ˢ or 3ˢ in sitting posture. No anæsthesia. 3ʰ 35ᵐ improvement commencing.	Recovered.	
Do. do.	1·8 grm.	42ᵐ cannot sit up. Hind legs almost powerless. Has a good deal of tremor in fore limbs. 1ʰ 50ᵐ pupils much dilated. Looks collapsed. Head tends to fall forwards. Heart and respiration regular. Never attempts to move. 2ʰ tends to roll on to side, but can still regain belly position. Shortly after died.	Died.	Next day. Brain appears rather congested, but cord does not. Intestines and stomach full. Right heart distended; left empty. No congestion of kidney. Slight congestion of liver.
8. Dimethyl-ammonium iodide	1·6–1 grm.	60ᵐ after injection violent shivering. Almost convulsive when movement is attempted. 70ᵐ attempts to remain sitting, only succeeds for a few seconds (1ˢ). 79ᵐ respiration feeble. Struggles now and then, otherwise quiet. 83ᵐ corneal reflex persists, but otherwise almost completely paralysed. 93ᵐ corneal reflex gone. Struggles. Death occurs within an hour from this time.	Died.	
9. Diethyl-ammonium iodide	1 grm.	In 50ᵐ appears weaker on legs, which are occasionally stretched out behind it. It is, however, able to sit up. 45ᵐ the stretching out of legs seems to be spasmodic in character, but this may be due to paralysis having advanced further in flexion than extension. The animal, however, still endeavours to restore leg to normal position. In 2ʰ 5ᵐ appears stronger. Able to sit up better. There has never been any anæsthesia.	Recovered.	
10. Trimethyl-ammonium iodide	2 grm.	First injected 1 grm. Second injection was 2ᵐ 45ᵐ after first. In 13ᵐ lies on belly. Hind legs specially weak. 16ᵐ getting stronger. 29ᵐ after second injection is apathetic, and trembles when taken up. Does not appear anæsthetic, and runs if placed on floor, and excited. Position normal.	Recovered.	
11. Triethyl-ammonium iodide	2 grm.	No symptoms but of weakness.		
12. Tetramethyl-ammonium iodide	1 grm.	Immediately shrieks. Salivation profuse. 5ᵐ lies apparently paralysed. 4ᵐ heart still beating. Cornea insensible.	Died.	Lungs not congested. Liver congested.
	·2 grm.	Recovered.	Recovered.	
	·5 grm.	15ˢ no shriek. Salivation profuse. 21ˢ head falls on one side. 4ᵐ respiration ceased. Heart beating. Movement of limbs very violent. 41ᵐ dead.	Died.	Liver congested. Brain and cord. Heart beat a long time after thorax opened.

ACTION of Simple and Compound Ammonias on Rabbits (continued).

Solution Injected	Quantity	Symptoms, with Notice of Time of their Occurrence in Brief	Ultimate Result	Post-Mortem Appearances.
14. Tetraethyl-ammonium iodide .	·3 grm.	In 11" animal trembles and shivers. Head tends to drop 20" head and forequarters much paralyzed. Corneal reflex still present. Reflex still in legs. 25" respiration ceases. Corneal reflex still good. Eye protruded. 29" nostrils still twitch, and heart beats. 36" dead.	Died.	No congestion of intestines. Heart still continued.
Do. do. .	·5 grm.	In 21" lies with head on table. Shaking from side to side. Head tends to fall on one side. 26" convulsive springing, harcheez, and shuffling. Pyagorus. Clonic spasm of jaw. 30" death.	Died.	Brain and cord; kidneys and liver congested.
15. Methyl-ammonium sulphate .	·7 grm.	No symptoms.	Recovered.	
	1·0 grm.	No symptoms.	Recovered.	
16. Ethyl-ammonium sulphate .	·7 grm.	Did not vary from normal, except in 25" the hind legs appeared rather weak.	Recovered.	
	2·0 grm.	No symptoms.	Recovered.	
16. Amyl-ammonium sulphate .	2·0 grm.	25" respiration accelerated. Legs slip from beneath it. 65" sinks on belly, but can rise again. 70" lies on belly with legs extended behind it, but can with take a few springs if excited. 95" effect of poison passing off.	Recovered.	
17. Dimethyl-ammonium sulphate .	·7 grm.	No symptoms.	Recovered.	
	2·0 grm.	In 10" sinks on belly on table, cannot rise. 62" starts when touched. 2° 55" begins to run.	Recovered.	
18. Diethyl-ammonium sulphate .	·7 grm.	In 35" lies with legs extended behind it, but can easily rise.	Recovered.	
	2·0 grm.	No symptoms.	Recovered.	
19. Trimethyl-ammonium sulphate .	·7 grm.	Per 2° 10" after completion of injection lies flat on belly.	Recovered.	
	2·0 grm.	In 12" tends to lie with legs out behind, but can sit up. Profuse salivation. 51" hind legs are quite paralyzed. Respiration snoring. Lies flat on side. 1° 25" raises head on noise. 3° 15" died. Corneal reflex to the last.	Died.	
20. Triethyl-ammonium sulphate .	·7 grm.	25" lies down on belly, then rises, but soon lies again. 1° 57" lies with legs extended behind it.	Recovered.	
	2·0 grm.	12" tremors in fore paws. Convulsive shudder. 17" reflex gone. Respiration ceases.	Died.	

Rabbits, Tabulation of Results.

Fatal.

10 c.c. or less.*	Fatal in—	20 c.c. or above 10 c.c.	Fatal in—
10 c.c. Dimethyl-ammon. chloride.	Several hours.	20 c.c. Triethyl-ammon. sulphate.	17ᵐ.
10 c.c. Trimethyl-ammon. chloride	A few hours.	20 c.c. Trimethyl-ammon. sulphate	3ʰ 15ˢ.
10 c.c. Diethyl-ammon. iodide. .	2ʰ 30ᵐ.	20 c.c. Ethyl-ammon. iodide . .	Several hours.
10 c.c. Tetraethyl-ammon. iodide .	2ʰ 30ᵐ.	18 c.c. Amyl-ammon. iodide. . .	3ʰ.
10 c.c. Tetramethyl-ammon. iodide	2ʰ 4ᵐ.		
5 c.c. Tetramethyl-ammon. iodide.	5ᵐ.		
5 c.c. Tetramethyl-ammon. iodide .	30ᵐ.		

Not Fatal.

	With Maximal Dose recovered from—		
2 c.c.	10 c.c. and more than 2 c.c.	20 c.c. and more than 10 c.c.
Methyl-ammon. sulphate.	7 c.c. Trimethyl-ammon. sulphate.	16 c.c. Triethyl-ammon. iodide.
	10 c.c. Ethyl-ammon. iodide.	20 c.c. Amyl-ammon. sulphate.
	10 c.c. Triethyl-ammon. iodide.	20 c.c. Ethyl-ammon. sulphate.
	10 c.c. Diethyl-ammon. chloride.	20 c.c. Diethyl-ammon. sulphate.
	10 c.c. Triethyl-ammon. chloride.	20 c.c. Dimethyl-ammon. sulphate.
	10 c.c. Amyl-ammon. iodide.	19 c.c. Methyl-ammon. sulphate.
	10 c.c. Diethyl-ammon. iodide.	20 c.c. Methyl-ammon. iodide.
		20 c.c. Dimethyl-ammon. iodide. .
		17 c.c. Trimethyl-ammon. iodide.
		17 c.c. Diethyl-ammon. chloride.

The order of fatality considering—

I. The salt.
    1. Iodides.
    2. Chlorides.
    3. Sulphates.

II. The ammonia compound.
    1. The tetraethyls and tetramethyls.
    2. The triethyls and trimethyls.
    3. The diethyls and dimethyls.
    4. The amyls, ethyls, and methyls.

appears to be :—

The former (I.) is of very secondary importance to the latter, and the difference between the iodides, and chlorides, and sulphates is magnified by the fact that in the case of the iodides alone were the *tetra* compounds employed.

In regard to rapidity of action, we find (1) tetramethyl-ammonium-iodide (5 c.c. = ·5 gr.) fatal in 5ᵐ; (2) triethyl-ammonium sulphate (20 c.c.) in 17ᵐ, and tetraethyl-ammonium-iodide (5 c.c.) in 30ᵐ. No symptom of pain occurred in any case after the injection, nor of physical change in animal. There appeared occasionally a slight loss of co-ordination, but this may have been, in some cases, due to paralysis. The pupil was markedly affected in the case of trimethyl-ammonium chloride and tetramethyl-ammonium-iodide.

    * Each c.c. is equal to ·1 grm. of the substance named.

From these experiments it appears that amongst the drugs contained in the table, the most marked disturbance occurs in the Rat in the case of the ethyl-ammonium sulphate, amyl-ammonium sulphate, amyl-ammonium iodide, diethyl-ammonium-sulphate, diethyl-ammonium chloride, triethyl-ammonium chloride, and tetra-methyla-mine ammonium iodide. In all of these tremors were noticed, and in some—the diethyl-, triethyl-, and amyl-ammonium salts—a peculiar rapping of the head upon the table was noticed, which appeared to be of a convulsive character.

The two most powerful convulsants were the amyl-ammonium-sulphate, and the tetramethyl-ammonium-iodide.

We found that the iodides not enumerated amongst those causing marked nervous disturbance have little tendency to produce spasmodic movements. In them loss of reflex, first in the hind legs, and then in the anterior part of the body, is most marked.

It appears to us that as a group the salts of the compound ammonias have a complex action: they affect the spinal cord, motor nerves, and muscles. The extent to which these structures are affected by the different compounds varies with each compound.

The spinal cord appears to be first stimulated, and then paralyzed. The symptoms which lead us to suppose that it is first stimulated are the twitchings which occur in the early stage in Rabbits and Rats, when poisoned with the substances mentioned in the tables, and the convulsions which occur in Frogs poisoned by ethylamine and tetraethylamine-iodide. That the spinal cord is paralyzed at a later stage, both as a conductor of motor stimuli and as a reflex centre, we infer from the failure of reflex action both in Frogs and Mammals, and from the fact that a stimulus applied to the hind foot frequently induces motion, not of the corresponding hind leg, but of one of the fore legs.

The convulsions which occur shortly before death in mammals are, perhaps, to be regarded as due, not to the irritant action of the poison on the nerve centres, but rather possibly to its paralyzing action on the motor nerves: this motor paralysis causes enfeebled breathing, and a consequent venous condition of the blood with asphyxial convulsions. That the compound ammonias and their salts paralyze the motor nerves is shown by our direct experiments on the nerve muscle preparation, in which the nerves were almost always paralyzed before the muscle. The muscles, however, are by no means unaffected—at first their power may seem to be increased, so that they respond by a more powerful contraction to irritation; afterwards, however, they become weakened, and tend to become completely paralyzed by the continued action of the poisons. This increase of irritability is not observed in the case of some of the compounds, even as a temporary condition.

COMPARISON BETWEEN THE ACTION OF AMMONIA AND THE COMPOUND AMMONIAS.

Ammonia itself has a convulsant action, the convulsions apparently being due to its effect upon the spinal cord, like those of strychnia. It differs, however, from strychnia in this respect, that the convulsions do not continue long, apparently because the motor nerves soon become exhausted, so that the excited spinal cord can no longer induce muscular contractions. The only one of the compound ammonias, in which one atom of hydrogen only is replaced by an alcohol radical, that we have experimented with is ethylamine; and this we find has also a convulsive action, probably the same in nature as that of ammonia. It has but a feeble paralyzing action on motor nerves. But this is only true of a single stimulus or of a few stimuli. When the nerve is subjected to rapidly repeated stimulation, it becomes very quickly exhausted. Ethylamine, therefore, while not directly paralyzing the excitability of the nerve, greatly lessens its endurance and power of work. It will thus have a similar effect to ammonia in shortening the convulsions, and thus rendering them like those of ammonia, and unlike those of strychnia. Its action on muscle itself appears to be very similar to that of ammonia. First it increases the excitability of the muscle, but afterwards diminishes it, and renders the curve both lower and longer.

Trimethylamine was found by HUSEMANN to have a tetanising action even on Frogs, like that of ammonia. In our experiments, however, we found gradual failure of the circulation and of reflex without any spasm. This difference between his results and ours may be possibly due either to our having employed different kinds of Frogs or to our having experimented at different seasons and under different temperatures. Another possibility is, that the Frogs he employed were stronger, and that their circulation was more vigorous than ours: for we have already noted that ammonium bromide produced tetanus in Frogs, but this came on at a late period in the poisoning, and unless the Frog was strong, and the circulation vigorous, the animal died before the tetanus made its appearance.

With triethylamine we noticed a great failure of reflex, unaccompanied by spasm; with both triethylamine and trimethylamine the action appeared to be slower than that of ethylamine. In one case of poisoning by the latter, tonic spasm occurred in 70ᵐ after injection, whilst in two hours after the injection of a larger quantity of triethylamine and trimethylamine, a faint reflex action was still present, and the circulation was maintained. The action of trimethylamine and triethylamine on motor nerves and muscle is very much the same as that of ethylamine or ammonia. From a comparison of ammonia with these compound ammonias it appears that the replacement of hydrogen by alcohol radicals tends to diminish the convulsant action of ammonia itself, and that the diminution is greater in proportion to the number of hydrogen atoms substituted.

We have not obtained any distinct evidence that the substitution of alcohol radicals for hydrogen increases the paralyzing action of ammonia on motor nerves, or indeed alters its effect upon the muscle.

We shall presently have to notice, however, the marked change which occurs in physiological action, when we pass from an ammonia in which nitrogen is combined with three atoms of an alcohol radical to those in which we have it combined with four atoms as tetramethyl- and tetraethyl-ammonium iodides.

CHLORIDES.

Ammonium chloride has been shown by BOEHM and LANGE to produce convulsions resembling ammonia itself.

Amylamine hydrochlorate has been shown by DUJARDIN-BEAUMETZ to have a convulsant action upon Rabbits.

Our experiments on Frogs have led to the following results :—

We found that methylamine chloride caused gradual failure of reflex action generally unaccompanied by spasm, while the diminished reflex produced by ethylamine chloride was of a spasmodic nature, though there was no true tetanus. With amylamine chloride we observed no spasm. In one case there was a tendency to spasm chiefly in the hyoglossus muscle.

The dimethyl- and diethyl-ammonium chloride cause weakness, lethargy, and failure of reflex action, but no distinct spasm. A tremor is observed on movement, but this seems to be rather due to failure of motor nerves than to increased excitability of nerve centres.

Their action upon motor nerves and muscles appears to have been much the same as that of ethylamine : the nerve not being directly paralyzed, but its power of transmitting stimuli continuously being greatly diminished.

The muscle has at first its contractility increased but afterwards diminished (Plate 8, fig. 1, a, b, c, d). In these experiments on Frogs the chlorine does not appear to have altered the action of the compound ammonias with which it is combined.

From experiments on Rats we find that both diethyl- and triethyl- ammonium chlorides have a similar action. The most marked symptoms are motor weakness and tremor. The tremor is most perceptible when the animal moves, and there is a very curious spasmodic movement of the head causing the chin to rap upon the floor.

Before death, convulsions occur, but these are probably asphyxial.

In Rabbits the effect is somewhat similar. The movements become tremulous, are exaggerated and scrambling in character, suggestive of impaired co-ordination. No anæsthesia was observed. Reflex was lost gradually and disappeared, first in the hind limbs.

The most marked effect of the chlorine in altering the action of the compound ammonias appears in these experiments to be a tendency to produce tremor. It is perhaps not quite easy to say positively what the cause of this tremor is, but we are inclined to regard it rather as an indication of failing power in motor nerves than to increased irritability in nerve centres.

IODIDES.

As we have already shown in an earlier part of this paper, ammonium iodide has a powerful paralyzing action, both on nerve centres and motor nerves, producing sluggish movements and motor paralysis.

From experiments on Frogs we find that methyl- (Plate 8, fig. 2, a, b), ethyl-, and amyl- (Plate 8, fig. 3, a, b, c) ammonium iodides all produce torpor. In the ethyl-ammonium iodide, GOLTZ's "croak" experiment succeeded as it did in the case of simple ammonia iodide. With the amyl-ammonium iodide, jerking or staccato movement of the limbs was observed, apparently due to failure of motor power. The methyl-, ethyl-, and amyl-ammonium iodides in small doses increase the excitability both of nerve and muscle. In large doses they are powerful poisons to motor nerves; they have a tendency to alter the formation of the muscle curve, and produce in it a curious hump, but they do not appear to affect muscle as much as nerve.

The occurrence of the croak in the ethyl-ammonium iodide would appear to indicate rapid paralysis of the higher nerve centres; and the staccato movement in the amyl-ammonium iodide, more rapid failure of motor nerves.

The dimethyl- and diethyl-ammonium iodides produced increasing lethargy, with no spasm ; with the diethyl-ammonium iodide the "croak" experiment succeeded, as it did with the ethyl-ammonium iodide.

Their action upon muscle and nerve seems to be similar to that of the methyl- and ethyl-ammonium iodides. Trimethyl- and triethyl-ammonium iodides have an action like that of the dimethyl- and diethyl-ammonium iodides, but they appear to have a greater paralyzing action on muscle and nerve (Plate 8, fig. 4, a, b, c), the primary increase in excitability not being marked, and paralysis of both occurring more readily. The tetramethyl- and tetraethyl- (Plate 8, fig. 5, a, b, c) ammonium iodides present a marked contrast to the other iodides, as Frogs poisoned by them exhibit spasmodic twitchings of the trunk and extremities. The higher reflexes cease very rapidly. The nerve is generally completely paralyzed. The muscle is only slightly affected when the poisoning is rapid, but if it be slow it is completely paralyzed also.

All the iodides render the beats of the heart slow, and tend to produce still-stand in diastole.

In the case of triethyl-ammonium iodide a vermicular movement of the heart was observed.

The tetraethyl- and tetramethyl-ammonium iodides appear to have a more powerful action than the others in producing diastolic still-stand of the heart.

Experiments on Rats.

Methyl-, ethyl-, and amyl-ammonium iodides all produce increasing weakness with a sprawling or waddling gait.   The power of the cord to conduct motor impulses appears to be diminished so that the hind legs become more paralyzed than the fore legs.   Its conducting power for sensory impressions is not paralyzed at this time, as stimulation of the hind legs will produce movement in the anterior part of the body. In the case of poisoning by amyl-ammonium iodide, twitching of the limbs and head were more marked than that of the methyl or ethyl compounds.

The dimethyl- and diethyl-ammonium compounds also cause progressive paralysis. In the case of the diethyl-ammonium iodide, an occasional instantaneous twitching in back and forelimbs was observed, resembling an effort at hiccough.

The tetramethyl-ammonium iodide has an action very different from the others, producing powerful convulsions.   It kills also much more rapidly, and is fatal in very much smaller dose.

Experiments on Rabbits.

In Rabbits the methyl-, ethyl-, and amyl-ammonium iodides all cause increasing weakness.   The conducting power of the cord appears here also to be affected, the hind legs becoming sooner paralyzed than the fore legs.

In the case of the methyl-ammonium iodide there is a distinct trembling of the body not noticed in the other two.

General Action of the Iodides.

A distinct alteration appears to be effected in the action of the compound ammonias by the combination with iodine.   All the iodides, both of ammonia itself and the compound ammonias, have a powerful paralyzing action on the motor nerves. Muscular irritability is as a rule decreased ; occasionally it is increased at first, as in the case of the methyl-, ethyl-, and amyl-ammonium iodides.

The muscle curve in all cases shows a tendency to become humped.   This tendency is more marked in the methyl, ethyl, and amyl compounds than in the di- or trimethyl, ethyl, and amyl compounds.   It is more marked when the muscle is stimulated directly than when it is stimulated through the nerve.   They all render the muscle more easily exhausted, so that the tetanic curve becomes lower and is sustained for a shorter time.

SULPHATES.

Experiments on Frogs.

Ammonium sulphate soon causes the movements to be accompanied with twitchings and clonic spasm. It sometimes, though rarely, produces complete tetanus ; the peripheral ends of motor nerves are paralyzed by it, and the muscular substance is also paralyzed, though later than the nerve.

The heart is considerably affected by the poison, and is frequently found arrested in diastole, and filled with dark blood. In this point it appears to agree with the iodide.

Methyl, ethyl, and amyl sulphates all cause gradually increasing lethargy and failure of reflex movement.

Methyl-ammonium sulphate paralyzes muscle and nerve very completely, the nerve being paralyzed before the muscle. The ethyl- and amyl-ammonium sulphates have much less paralyzing action upon muscle and nerve, but render them liable to rapid exhaustion.

In poisoning by them the heart was considerably affected, and beat very slowly ; probably the slighter effect on the muscle of ethyl and amyl sulphates in our experiments was due to their greater effect upon the heart, so that they were carried in lesser quantity to the muscle. This is exactly what one finds with such a poison as ventrine, which has an extraordinary effect on the muscle of a Frog in small doses, but has little effect on the muscle when the dose is large, the heart being so quickly arrested that but little effect is produced upon the muscle.

Dimethyl- and diethyl-ammonium sulphate both cause weakness, with tremulous movement ; but in the case of diethyl-ammonium sulphate, strong irritation causes a powerful movement in the limbs, occurring after a considerable latent period. The nerve appears to be powerfully paralyzed by the dimethyl-ammonium sulphate, while the paralyzing action is but slightly marked in the case of the diethyl-ammonium sulphate ; the paralysis of the muscular tissue is also more marked in the case of the dimethyl-ammonium sulphate (Plate 8, fig. 6, *a, b*). Both lessen the activity of the circulation, and render the cardiac pulsations slow.

The trimethyl- and triethyl-ammonium sulphates both cause the movements to become weaker and tremulous, and sometimes staccato.

The trimethyl-ammonium sulphate (Plate 8, fig. 7, *a, b, c, d*) appears at first to increase the excitability of the animal, and even when the muscular power has failed, so that irritation of the foot no longer will cause it to be withdrawn, tremor occurs over the whole body from the stimulus. The nerve is either much weakened or paralyzed, so that it either soon gives way when tetanised, or does not respond to stimulus at all. The muscle is also paralyzed ; the minimal irritability is much impaired in poisoning by trimethyl-ammonium sulphate, although the contractile power remains considerable.

One of the most marked points in the action of the sulphates of ammonia and compound ammonias on the Frog appears to be their tendency to affect the circulation, and to render the beat of the heart slow, or arrest it entirely in diastole. Muscle and nerve are both paralyzed, the paralysis of the muscle being later than that of the nerve.

We have noted above a number of more or less exceptional instances, but in many of those there can be little doubt, we think, that the exceptional action was due to alteration in the circulation caused by the poison.

In their action upon the circulation the sulphates resemble the iodides. The spinal cord appears to be stimulated, so that convulsions or tetanus are produced by the ammonium sulphate. The combination with ethyl and methyl appears to lessen this stimulating action, although we notice in the case of the triethyl-ammonium sulphate a tendency to diffusion of stimuli in the cord, irritation of the foot being responded to by tremor over the body.

In the case of the Rat we find the amyl-ammonium sulphate to be one of the most poisonous of the whole series used in the case of these animals. There is violent tremor, increased on movement; a gait like that of paralysis agitans; sudden general clonic spasm, succeeded by springing from side to side.

In the case of ethyl-ammonium sulphate and diethyl-ammonium sulphate the movements are likewise tremulous; rapping of the head upon the floor is observed, and there is frequently a spasm of many of the trunk muscles, giving the impression of a hiccough movement. Respiration, at first accelerated, becomes very feeble, and a gradual loss of reflex precedes death.

The circulation was slowed by the action of these poisons, the heart tending to diastolic arrest, the right side especially being much engorged.

It was found that stimulation, both direct and indirect, elicited a powerful contraction of the poisoned muscle. The changes in circulation no doubt account for the slight effect of the poison upon the muscle. In the case of the amyl-ammonium sulphate, congestion of the membranes of the brain and of the cord itself were observed.

General action on Rabbits.

In the case of Rabbits, in which the whole series of these poisons was investigated, there was observed a gradual loss of power, the animal tending to lie on the belly, with the legs extended; the hind legs appeared to be chiefly affected.

In the case of the triethyl-ammonium sulphate, and the trimethyl-ammonium sulphate, there was a certain amount of tremulousness and starting when touched. The paralysis in the hind legs became complete before it did in the fore legs.

In the case of trimethyl-ammonium sulphate, profuse salivation was an early symptom, and corneal reflex persisted to the last.

The sulphates were less fatal to Rabbits than the corresponding chlorides or iodides, with the exception of trimethyl and triethyl sulphates, in which there was trembling and slight spasmodic movements, probably indicative of irritation of the spinal cord.

The symptoms were those of paralysis of the spinal cord and motor nerves. The conducting power of the cord for motor impressions appears to be paralyzed, as the hind legs fail before the fore legs. Death occurs in Rabbits and Rats by failure of respiration.

DIFFERENCE BETWEEN THE ACTION OF THE SALTS OF THE COMPOUND AMMONIAS.

Our experiments appear to us to show that the salts of the compound ammonias vary in their action : (a) according to the acid radical with which they are combined ; and (b) according to the number of the atoms of hydrogen which have been replaced in the ammonia by an alcohol radical. The influence of the acid, however, appears to us to be less marked than in the case of ammonia itself.

The iodides appear to have the strongest paralyzing action, both on the central nervous system and on the peripheral nerves. Next to them come the chlorides, and the sulphates have the least action.

The paralysis of the higher reflex, e.g., of the cornea, was more marked in Frogs than in Mammals. In the latter, indeed, corneal reflex was observed almost at the last.

We have only examined the action of the iodides of tetramethyl- and tetraethyl-ammonium, so that we cannot compare their actions with those of the corresponding chlorides and sulphates. We have already drawn attention to the fact that their action appears to differ very greatly from the compound ammonias in which only three atoms of hydrogen have been replaced by an alcohol radical. In the tetra compounds convulsant action is very strongly marked, while in the triad compound ammonias it is much less so, or may be altogether absent.

In the case of warm-blooded animals salivation was noticed before death in poisoning by trimethyl-ammonium sulphate, tetramethyl-ammonium iodide, and tetra ethyl-ammonium iodide ; it also occurred, to some extent, in amyl-ammonium iodide. In one or two others a similar action was observed to a less extent.

We have not investigated fully the action on the spinal cord and higher nerve centres of these different compounds, because the number of substances on which we have experimented was so great that we thought it better to leave this subject for a subsequent research, and to confine ourselves more especially to their action on muscle and nerve.

The results of our experiments on these tissues are shown in a condensed form in the following paragraphs :—

DIFFERENCES BETWEEN THE ACTION OF SALTS OF THE COMPOUND AMMONIAS ON THE FROG'S MUSCLE AND NERVE.

For convenience sake we will group the bodies, first, according to the acid radical ; and secondly, according to the base they contain.

VARIATIONS IN ACTION ACCORDING TO THE ACID RADICAL.

Chlorides.

(*a.*) *Irritability* is, as a rule, slightly increased.

(*b.*) *Tetanus* from the muscle is often more extensive, whilst that from indirect stimulation is less extensive than on the normal side.

(*c.*) The *curve* is often exaggerated in direct stimulation.

It is frequently higher, and may be slightly shorter or longer than normal. On repeated stimulation, whether direct or indirect, the curve elongates to a greater or less extent. There is, as a rule, less elongation, less succeeding contraction, and less tendency to develop a distinct second hump than is to be seen in the iodides.

(*d.*) The nerve gives way somewhat before the muscle, but these substances (*i.e.*, chlorides) are not so fatal to nervous irritability as are the iodides. Amyl-ammonium chloride has a relatively stronger action on nerve than on muscle.

Iodides.

(*a.*) *Irritability* is, as a rule, decreased, the exception being occasionally found in ethyl-ammonium iodide, and di- and triethyl-ammonium iodides.

(*b.*) *Tetanus* is diminished in extent in almost every case.

(*c.*) The *curve* shows a strong inclination in all, but most in those lowest in the series, to become two-humped, the second horn or hump passing into a contracture, with very gradual decline.

(*d.*) In all cases the nerve becomes paralyzed much before the muscle.

Sulphates.

(*a.*) Minimal *irritability* is increased, or normal in the case of ethyl-ammonium sulphate, diethyl-ammonium sulphate, and triethyl-ammonium sulphate. It is decreased by amyl-ammonium sulphate, and by all the methyl-sulphates.

(*b.*) *Tetanus* produces more extensive contraction on direct stimulation in the case of the ethyls, and in very slight poisoning in some instances in the methyls, but in the latter it is usually diminished.

(*c.*) The *curve* is chiefly affected by the methyl compounds, on which it is usually lower and longer, and shows increased viscosity. It seldom displays the strong tendency to the double hump form which is so common amongst the iodides.

In the ethyl compounds the curve is usually somewhat exaggerated in relationship to the normal.

(*d.*) The failure of the nerve occurs somewhat sooner than that of the muscle. This is much more marked in the methyl than in the ethyl compounds.

On summing up these results, it appears that the iodides paralyze motor nerves more quickly than either chlorides or sulphates. We did not observe any marked difference between the paralyzing action of the corresponding chlorides and sulphates. In the case of the muscle we notice that the irritability is increased, as a rule, in poisoning by the chlorides; is sometimes increased and sometimes diminished by the sulphates; and, as a rule, though with some exceptions, it is decreased by the iodides. The contractile power of the muscle, as shown by the extent and duration of tetanic contraction on direct stimulation, appears to be least affected by the chlorides; somewhat more so by the sulphates; and most of all by the iodides. The alterations in the form of the curve have already been described in detail.

VARIATIONS AMONGST THE ETHYLS AND METHYLS.

The least operative compounds examined were the diethyls and triethyls. Thus, in these alone, in the case of the iodides and sulphates, was minimal irritability equal to or greater than the normal.

(a.) In the case of the chlorides, however (in which the ethyls, methyls, di- and trimethyls only were examined), there was not a material difference between the corresponding compounds.

(b.) Amongst the iodides there is a strong tendency to loss of irritability of the nerve with all the compounds, but this is pre-eminently the case with the tetraethyl- and tetramethyl-ammonium iodides, which have an extremely powerful paralyzing action. The methyl compounds appear, however, to be operative in a slightly smaller dose.

(c.) The smaller group of the chlorides does not present such striking variations, but the corresponding methyls are slightly more active than the ethyls.

(d.) Amongst the sulphates we find the ethyls more often to produce an exaggerated single curve and an increased tetanus than do the methyls. There may, however, as shown in the chart of trimethyl-ammonium sulphate, be an increase in tetanic contraction as a result of stimulation in an early stage of poisoning.

(e.) The methyl compounds of the sulphate group are decidedly more fatal to the irritability of the nerve than are those of the ethyls.

(f.) *Ethylamine* showed development of *tetanic spasms* 70″ after injection. There was a gradual failure of reflex and circulation.

There was increased irritability to both direct and indirect stimulation; the curve was higher, longer, and showed increased viscosity.

Triethylamine—gradual failure of reflex and of circulation. Increased viscosity of the muscle was observed, without a marked lengthening of the curve.

Trimethylamine—gradual failure of reflex and of circulation. Increased irritability and increase of viscosity. The curve is equal to or shorter than the normal.

The methyls are more active than the corresponding ethyls. The methyls, amyls, and ethyls are more effective than the corresponding di- and tri- compounds. The tetra compounds are, however, most so of all.

ACTION OF SALTS OF THE ALKALINE GROUP ON MUSCLE AND NERVE, AND A
COMPARISON OF THEIR ACTIONS WITH THAT OF AMMONIA.

The bodies usually included in the group of alkalies are, in addition to ammonia, lithium, sodium, potassium, rubidium, and cæsium: these are all monads. MENDELEJEFF includes in the monad group copper, silver, and gold, in addition to the substances just mentioned; but there is such a well marked difference between the general properties of the metals last mentioned and those of the alkalies that we have not included them in our research.

On comparing the general action of ammonia with these substances, the first thing that strikes us is that ammonia is the only one which has any tetanising action. Sometimes reflex action seems to be a little excited at first in poisoning by potassium and rubidium, but this excitement is slight, soon passes off, and is succeeded by torpor.

In the case of sodium, lithium, and cæsium, the symptoms in Frogs are those of gradually increasing torpor.

Sodium has no action at all in small quantities, but in concentrated solutions appears to paralyze nerve centres, nerves, and muscles, all at the same time. Lithium, rubidium, and cæsium have a tendency to affect either the upper part of the spinal cord or the higher motor centres connected with the fore limbs, as in poisoning by lithium and cæsium the reflex disappears sooner from the arms than from the legs, and stiffness was noticed in the arms in poisoning by lithium and cæsium, though no distinct spasm was observed. The motor nerves are not paralyzed by sodium or rubidium, but with these exceptions they are paralyzed to a greater or less extent by the other substances belonging to this group. Lithium and potassium are most powerful.

In considering the effect of the alkalies, and still more, perhaps, in the case of the alkaline earths, we have carefully to distinguish between the action of the poisons on the active contraction of muscle and on the residual shortening, which continues for a greater or less time after the contraction has passed.

To this shortening we have sometimes given the name of *viscosity*, at others, and more generally, we have employed the term used by German and French writers, *contracture*.

In regard to active muscular contraction also, we must note both the height of the curve, indicating the amount of contraction and its length, indicating the length or duration of contraction. The exact difference between the action of the various substances will be seen more in detail by a glance at the accompanying tables and curves.

But we may here state generally that the contractile power of the muscle, as shown by the height of the curve it describes, is increased by ammonium, potassium, and sometimes by rubidium and cæsium.

It is occasionally increased by sodium, but is otherwise unaffected, excepting in large doses, and it is diminished almost invariably by lithium.

The duration of the contraction, as shown by the length of the curve, is increased by large doses of rubidium (Plate 8, fig. 8, *a*, *b*, *c*), ammonium (Plate 8, fig. 9, *a*, *b*), sodium (Plate 8, fig. 10, *a*, *b*, *c*), and cæsium (Plate 8, fig. 11, *a*, *b*). It is shortened by ammonium (Plate 8, fig. 12, *a*, *b*), lithium (Plate 8, fig. 13, *a*, *b*), rubidium, and potassium (Plate 8, fig. 14, *a*, *b*, *c*). It will be seen from this enumeration that rubidium, ammonium, and sodium have a double action, sometimes increasing and sometimes diminishing the length of the contraction. In the case of rubidium and sodium the difference of action depends upon a difference of dose, small quantities tending to shorten the contraction, while large doses lengthen it. Prolonged contraction is accompanied, as we have already mentioned, by an increase of contractility in the case of rubidium, but by a diminution in the case of sodium, as shown by the height of the curve. The double action of ammonia does not seem to us to depend entirely on difference of dose, but rather to the ammonium having two different kinds of action.

The residual shortening, viscosity, or *contracture*, which sometimes succeeds an active contraction, is increased by large doses of rubidium, ammonium, lithium, and sodium. It is diminished by rubidium in small doses, ammonium, cæsium, and potassium. Here, again, the different action of ammonia does not appear to us to depend entirely on difference of dose.

Its double action appears to form, to a certain extent, a connecting link between the action of some members of the alkali group, such as potassium, and that of members of the group of alkaline earths.

The relations between the various members of the present group have to be considered more fully in a subsequent section, because we find that some members of it, while having a somewhat similar action on normal muscle, will yet antagonise each other's action, and although either of them given alone will lengthen the muscular curve, the lengthening will be abolished, and the curve reduced to the normal, by the administration of the two together.

ACTION OF SUBSTANCES BELONGING TO THE GROUP OF ALKALINE EARTHS AND EARTHS.

The metals which we have examined belonging to the group of alkaline earths are calcium, strontium, and barium; and to that of the earths beryllium, yttrium, didymium, erbium, and lanthanum. The first three are dyads. Beryllium is also a dyad. The atomicity of the last four is not determined. Possibly they are all triads, though lanthanum has been grouped by MENDELEJEFF amongst the tetrads. The first point of difference that we notice about this large group is that it may be subdivided into two sub-groups:—(a) containing beryllium, calcium, strontium, and barium; and (b) containing yttrium, didymium, erbium, and lanthanum.

In group (a) we notice a tendency to increased reflex action. In this particular it agrees with ammonium, but differs from members of the alkaline group. We have already noted that, in some members of the alkaline group, a slightly increased reflex action might be observed at the commencement of the poisoning, but this is considerably less than in the case of most of the members of group (a), with the exception of barium. Excitement of the spinal cord is most marked in poisoning by beryllium; next come strontium and calcium; and lastly barium, in which excitement, if present at all, is very slight.

In group (b) reflex action in the cord is not increased, nor does it appear to be very much diminished till the last. In this group, however, the higher centres appear to be paralyzed. We infer this from the fact that yttrium greatly diminishes co-ordinating power in the Frog, rendering the movements ataxic, and causing the animal to lie with the legs fully stretched out, although neither muscle or nerve is paralyzed. Didymium, erbium, and lanthanum all have a similar action.

In regard to their action on motor nerves, we notice the same well marked division into two groups as in their general action: beryllium, calcium, strontium, and barium all paralyzing the motor nerves to some extent. Lanthanum has also a paralyzing action, but yttrium, didymium, and erbium have none. In this respect these three bodies agree with sodium and rubidium, and differ from all the others belonging to these two groups which we have examined.

In regard to their action upon muscle, we do not find that these bodies can be so readily subdivided into two well marked sub-groups.

The contractility of muscle, as shown by the height of the curve, is greatly increased by barium (Plate 8, fig. 15, a–d), and occasionally, to a small extent, by erbium (Plate 8, fig. 16, a, b) and lanthanum (Plate 8, fig. 17, a, b). It is sometimes increased and sometimes diminished by yttrium (Plate 8, fig. 18, a, b) and calcium (Plate 8, fig. 19, a, b, c). It is diminished by didymium (Plate 9, fig. 20, a, b), strontium (Plate 9, fig. 21, a, b, c), and beryllium (Plate 9, fig. 22, a, b; fig. 23, a, b). We have found that the small variations occurring in the extent of contraction are best observed when the poison is applied locally in the form of solution. Where the muscles have been examined of an animal completely poisoned with the substance, the ultimate, rather than the primary result, is obtained.

The duration of the contraction, as shown by the length of the curve, is increased by barium, calcium, strontium, yttrium, and erbium. It is unaffected, or slightly diminished, by beryllium, didymium, and lanthanum (see figs. 17, 20, 22). It is obvious that the action of the rarer metals beryllium, erbium, didymium, lanthanum, and yttrium is but feeble in any direction when compared with the effect of calcium, &c.

The contracture is increased by barium, calcium, strontium, yttrium, and beryllium.

Contracture produced by barium is enormous (Plate 9, fig. 24, *a-g*). When the drug is locally applied its curve resembles greatly that produced by veratria (Plate 9, fig. 24, *b*). It appears to us to be an interesting fact that an inorganic element and an organic alkaloid should have such a similar action. Their action coincides also in the modifications which it undergoes by heat and by potash. The barium contracture, like that caused by veratria, is abolished by cooling the muscle down, or by heating it considerably above the normal. The contracture may be permanently removed by cooling down, so that it does not return when the muscle is again raised to the normal temperature. Like the veratria contracture, however, it is abolished much more certainly by heat (Plate 9, fig. 24). There is a more marked tendency for the barium contracture to relax suddenly than that caused by veratria. It is also more easily abolished by repeated stimulation.

In regard to the effect of these drugs on contracture, the same differences are to be observed between their action when injected into the circulation and when locally applied that we have already mentioned in regard to the active curve. In the accompanying diagram we have arranged some of the more important substances

belonging to the alkalies and alkaline earths so as to show their action upon muscle graphically. It will be seen that they tend to form a series, the two ends of which present some points of approximation, ammonium appearing to form a connecting link between barium and potassium.

It will be noticed that the substances here do not arrange themselves according to their atomic weight, nor yet according to their atomicities. We hope, however, to be able to consider this point more fully at a future time. We subjoin a table showing the relative position of the elements in regard to their action on motor nerves and muscles.

TABLE showing the relations of the Alkalies and Alkaline Earths as Poisons to Nerve and Muscle.

The most powerful paralyzers of motor nerves are put at the head of the column, and the others follow in the order of decreasing activity.

Those bodies which increase most the height and duration of muscular contraction and of muscular contracture are placed at the head of the corresponding columns, and at the foot are those which reduce them most.

Motor nerves.	Muscle.		
	Height of contraction.	Duration of contraction.	Contracture.
NH_4	Ba	Ba	Ba
L	Rb	Rb	Rb
K	NH_4	NH_4	NH_4
Be	Er	Na	Na
Ca	K	Ca	Ca
Sr	Cs	Sr	Sr
Ba	La	Yt	L
Cs	Yt	Cs	Yt
La	—	Er	Be
—	Ca	—	Di
Er	Na	Be	Er
Di	—	Di	Rb
Yt	Di	La	NH_4
Rb	Sr	—	Cs
Na	Be	NH_4	La
	L	L	K
		Rb	
		Na	
		K	

TABLE showing the relations of the Alkalies and Alkaline Earths as Poisons to Nerve and Muscle—continued.

Substance.	Animal.	Dose or Application.	Symptoms in brief.	Post-Mortem Appearances.	Reaction.	Antagonism.
Ammonium chloride	Frog	Local application— 1–1000 ... 1–500 1–200	...	...	Lengthens and heightens curve, increasing also the genuine shortening during the application of the solution. There is little or no increase of after-action (contraction). With weaker solutions may shorten curve. Lengthens (often greatly) and usually lowers curve, increases contraction.	The curve is markedly shortened by potash. And after potash has been long applied and the curve has become feeble, ammonium chloride restrains acidity of muscle. Lime shortens active curve, increases its altitude, and develops its after-action. Restores irritability lost by barium application.
Potassium chloride	Frog (15 grms.) (30 grms.)	5 (at once) 20 (gradually) [all once], all reduced gone.	In rapid poisoning (5%). Crouching attitude. All reflex gone, but has drawn up occasionally spasmodically. Circulation ceased. Fingers-tissue contracted. In slower poisoning, spring sudden and feebly. Tends to sink on belly. Leg moved side of body in snake manner. Galvanism of limb-muscles acts post-injection. Circulation-ceased still last injection. The respiration was hurried.	Heart in diastole, full of dark blood, no longer irritable.	In rapid (immediate) poisoning. Minimal irritability decreased (very slightly) on poisoned side. Tetanus from (direct) muscle stimulus. Curve slightly higher. Shorter with more rapid relaxation. In slower poisoning, there is a well-sustained tetanus which hardly causes a visible contraction however. The nerve is completely irritable.	Counteracts urtaria, barium, calcium, strontium. Lengthens muscle, raw curtails (barium, &c.). Reduces contraction of strong salt-solution muscle. Reduces contraction of lithium.
Caesium chloride	Frog (14 grms.) (30 grms.)	40 grm. ... Local application— body 1–1000	Gradually decreasing weakness. Reflexes died good, but more difficult to excite. Tendon knee-reflex, withstood a long time later than in snaps, arms may to sink. In 15 all reflex gone. Never any spasms. Convulsions good throughout, active even when reflex ceased. Tyanoic-cells uninjected.	Heart beating	At first and in weaker solutions shortens (hastens beginning and slight) lengthens curve. Prolonged action of weak solution or shorter action of stronger solution (1–400) lowers curve. Minimal irritability about equal in both muscles (viz., diminished in poisoned muscle?). Poisoned muscle is less irritable to breaking, current. On stronger stimulation a good tetanus (direct stimulation) as yet, but the nerve tetanus is very feeble, and there may be contraction only after exposing current. The curve is at first slightly increased in altitude, rapid in after clock. It is longer than normal, but only slightly so. In snow excursion poisoning the curve is rounded, this indicates an excited rapid time in the nerve.	Potash appears to shorten the elongated curve which may reduce in curtains. In any case lime did not cause any recovery.
Sodium chloride	Frog (15 grms.) (30 grms.)	7 grm. 7 grm. Local application— 4 per cent. 7 to 1 per cent.	Movements become gradually feebler, and more difficult to provoke. Tardily. Reflex become feeble, and in them completely lost. Circulation perfects refers. Many leucocytes seen in web, some migrating. Red corpuscles fewer and crenated.	Auricles sensitive to stimulation; ventricle less so. Its moderate distension.	In both cases of extreme potential, irritability is about equal (i.e., diminished in both muscle), the tetanus both direct and indirect is equal, well sustained, but only fair as stimulation on the normal. The curve is of a low and very prolonged, especially so in the case of direct stimulation. Applied locally of the strength of 7 per cent., there appears to be no active change in the curve. (This is therefore rightly called normal salt-solution.) 4 to 10. There is often a shortening (slight) of the active curve, and it is usually without any increase in altitude taking place. 1 to 17 per cent. There may as first be a slight shortening of the curve, but there is from the first an increase in altitude, which is accompanied by a lower (rapidly falling) curve, with considerable after action. The after-action may then become magnetised into a tetanus. 2 per cent. often reduces the curve in length at once, but also much in altitude, the muscle dying in short time.	These actions, which are strong enough to reduce another curve are themselves slow in action. Solutions of less than 2 per cent., or occasionally 1½ per cent., increase the altitude and quicken action. They also reduce height of curve action of calcium. Where there is, as usual, calcium, salt action developed potash opposes soda. There is partial opposition to the action of strontium soda, without a calculable curve developed. Potash often cuts off the contractions of strontium soda without in opposing to contraction.

* The irritability of the new-coloured muscle is distributed by the heater.

TABLE showing the relations of the Alkalies and Alkaline Earths as Poisons to Nerve and Muscle—continued.

Substance.	Animal.	Dose or Application.	Symptoms in brief.	Post-Mortem Appearance.	Reaction.	Antagonisms.
Lithium chloride	Frog . . . . (16 grms.) (13 grms.)	·03 grm. . . . ·04 grm.	Quietness. Spring weaker and shorter. Reflex persists longer in hind than fore legs. Reflex entirely lost in 1h to 2h or less. No tremor or spasm. Circulation steady at first, but becomes feebler. May be existing when reflex has ceased. Pigment cells generally contracted.	Heart often beating feebly.	There is generally no reaction whatever on stimulating the nerve, even by tetanising current, and whilst the muscle gives a good single contraction. The curve from direct stimulation is lower, rounder, longer, and with a very gradual relaxation, or the relaxation may be more normal. The tetanus from direct stimulation is not so extensive as in the unpoisoned muscle, but it is often well-maintained and considerable.	Potash heightens and takes away treble after-action. Soda (·75 per cent.) increases tonus without heightening curve. Lime heightens curve and increases tonus, and is again increased in tonus by lithia. Potash reduces both.
		Local application— 1-150 1-500	. . . .	. . . .	It is doubtful whether the normal curve is really shortened at all by lithium. At first the summit may be rather reduced, and hence an apparent shortening results; but this soon gives place to a lengthening. The passage of the curve active into a tonus is most marked and invariable in this curve. The altitude is never increased above the normal.	
Rubidium chloride	Frog . . . . (10 grms.)	·015 grm. .	Alternating restlessness and quiet. Leg drawn up quickly and high (power increased). Some stiffness of fore limbs. Movements become weaker and staccato. Reflex becomes feebler and unsustained, though rapid. At time of examination a spontaneous kick sometimes occurred. Breathing accelerated in early stage. Circulation, for some time good, gradually lessened in web, and ceased before heart stopped.	Heart in diastole, but began contracting again when blood was let out.	Irritability of both nerve and muscle to minimal stimulation diminished. The curve of the rubidium muscle, both for direct and indirect stimulation, is rather higher and rounder in the summit, with more rapid relaxation, than the normal, and therefore slightly shorter—or, at any rate, no longer—than the normal. Tetanus of both direct and indirect stimulation is more excessive.	Rubidium is antagonised both by barium and calcium. The curve is not altered by potash, except when veratria-like curve is produced, in which case potash reduces it.
	20 grms. . . .	·1 grm.	Arm reflex ceased first. Leg reflex ceased 45m after injection.	Heart in great diastole, full of dark blood.	Curve from both direct and indirect stimulation good.	
	16 grms. . . .	·02 grm.	Reflex diminishes. Leg drawn up weakly and tremulously, or remains extended. Heart ceased.	Ditto ditto, not irritable.	A low, long contraction from both direct and indirect stimulation.	
		Local application— 1-400 1-100	. . . .	. . . .	With weaker dilutions the curve is usually shortened, and sometimes heightened. After the action of the solution has continued some time the curve becomes lower and flatter. 1-100. Not unfrequently there is a distinct veratria-like curve, or there may be a considerable contraction after active curve instead.	
Calcium chloride .	Frog . . . . (14 grms.) (16 grms.)	·33 grm. . ·25 grm.	Primary exaggeration of reflex. Reflex gradually lost. Occurs last in legs. No tremor. Heart beats slowly (12 per 1m). Circulation in web gradually ceases. Pigment-cells contracted to balls. Formation of cataract in lens.	Heart ceased. Irritable.	Minimal irritability increased on poisoned side by 5 to 10 centims. The curve is usually higher for direct and sometimes for indirect stimulation. The curve is longer, and shows considerable "after action." Tetanus of muscle slightly inferior to the normal; that of nerve decidedly less, and yielding rapidly.	Is antagonised by potash. Reduced by very strong soda (1 to 15 and 1 to 60). Increased by strontium.
		Local application— 1-100 1-400	. . . .	. . . .	With weaker solutions there is often an increase in extent of contraction, diminution in extensibility, slight lengthening of the area, and increase of after-action. With stronger solutions there is a lowering (flattening) of curve, which is also lengthened, and there is an increase in after shortening, which is very marked.	

TABLE showing the relations of the Alkalies and Alkaline Earths as Poisons to Nerve and Muscle—continued.

Substance.	Animal.	Dose or Application.	Symptoms in brief.	Post-Mortem Appearances.	Reactions.	Antagonism.
Strontium chloride	Frog	·11 grm.–·960 grm.	Slight increase of reflex irritability.	Heart ceased. Ventricle only slightly irritable.	Minimal irritability direct and indirect stimulation.	Is antagonised by potash.
Barium chloride	Frog	·08	Position stiff.	Articles beating.	Motor irritability of barium nerve increased.	Is antagonised by:—Potash, Sodium, Calcium, Strontium, Lithium, &c., Rubidium.
Beryllium chloride	Frog	Injection ·22	Before may be increased soon after injection.	Heart often beating feebly.	There is usually complete paralysis of the nerve and muscle.	It increases altitude of tetanic curve. Calcium does not.

TABLE showing the relations of the Alkalies and Alkaline Earths as Poisons to Nerve and Muscle—continued.

Substance	Animal	Dose or Application	Symptoms in Intact.	Post-Mortem Appearance.	Reactions.	Antagonism.
Polyconium chloride	Frog (2.5 grm.)	Injection, .075 grm.	Became difficult in many to movement;—before being still good. No marked tremor. In plastic, fine to unnatural positions. Reflex becomes more feeble and slow. Circulation at first good, gradually ceases.	Heart in unirritated in moderate diastole, but goes on again no stimulating.	Minimal irritability {Sti. part. {N., 600 contract. "" "" {M., 125 "" (Lep. part. {M., 125 "" {M., 108 Tetanus (direct and indirect) less stimulation than normal {N., 40 "" {N., 42 }Normal. {M., 42 {M., 45 Curve (direct and indirect) to me quite so high, but is otherwise not abnormal. Does not materially alter length of curve, simply reduces altitude somewhat without materially altering form.	K. cl. 1·020 lengthens and heightens diphasma curve.
Thium chloride	Frog (.04 grm.) (.05 grm.)	Injection, .04 grm. "" .05 grm.	At first no motor variation, then a very gradual deterioration of reflex. Tendency to plasticity and emaciation of general positions. In one case passing frequent. Circulation for long time active, and pigments cells dilated; then circulation becomes feebler.	Heart beating feebly, sootzker stronger than the ventricle.	Increase contracture in resting muscle. Minimal irritability very equal. Tetanus equal. Curve of poisoned muscle higher (and longer from wash)—this rise in part in the long ligature of the other limb.	
Lavidanum chlor.	Frog (.05 grm.)	.05 grm. (Frog. 52 grm.)	Lies flat. Arms extended. Arms stiff. Will then lie with leg extended. Reflex good throughout.		Three to five a slight shortening and elevation of the normal curve. The curve soon becomes reduced in altitude. There is no increase of contracture.	Potash lengthens the length am curve. Calcium markedly heightens, lengthens, and causes after-action at end of contraction.
		Local application 1·100				
Yttrium chloride	Frog (.25 grm.) (.05 grm.)	.05 grm. ""	Restlessness succeeded by sitting up in stiff and unusual position. No tremor. Lethargic. Leg draws up wide of body in an unnatural motion. Legs become a little extended. Breathing accelerated. Circulation good throughout.	Heart beating	Minimal stimulation, irritability of lanthanum never greater than normal. Electrotonus (direct) is equal, (indirect) lanthanum muscle somewhat impaired. Curve of lanthanum (direct and indirect) somewhat shorter and higher—shorter because of more rapid reduction. Causes increasing tonus (passive contraction); a slight shortening of curve—more apparent than real, owing to the rise and shortening, which continually raises abscissa. Tetanus equal, or rather stronger in the yttrium gastrocnemius close to the long ligature of the contract leg, 30 to 80. Curve (normal) equal to or slightly less in altitude than the plateau. Yttrium curve has very slow relaxation in its lower part, giving it quite a special appearance. This is equally the case in direct or indirect stimulation. Irritability is greater than in the normal. Slight passive shortening of muscle. Slow relaxation after contraction. There is, however, shortening of more active part of curve in some cases.	
		Local application 1·100				

The effect of heat, cold, and stimulation must be postponed to another paper.

ACTION of elements upon general condition of organism as a poison acting gradually.

Substance.	Proportion to gramme of body-weight of Frog in which element acts as a poison.
Potassium chloride	·0013
Beryllium choride	·0013
Rubidium chloride	·0013 to ·0015
Barium chloride	·0013
Ammonium chloride	·0015
Cæsium chloride	·0021
Lithium cloride	·0023 to ·0032
Lanthanum chloride	·004
Didymium chloride	·0042
Erbium chloride	·006
Strontium chloride	·0055 to ·0075
Yttrium chloride	·009
Sodium	·0095
Calcium	·013 to ·017

Monads.		Atomic weights.	Dyads.		Atomic weights.
Potassium	·0013	39·10	Beryllium	·0013	9·4
Rubidium	·013 to 15	85·4	Barium	·0013	13·7
Cæsium	·0021	13·3	Lanthanum	·004	93·6
Lithium	·0023 to 32	7	Didymium	·004	95
Sodium	·0095	23	Erbium	·006	112·6
			Strontium	·0065	87·6
			Yttrium	·009	61·7
			Calcium	·013	40

ON THE ACTION OF ALKALI AND ACID ON MUSCLE.[*]

The remarkable results obtained by GASKELL upon the action of very dilute acids and alkalies on the blood-vessels, induced us to examine the action of similar solutions upon voluntary muscle. GASKELL found that alkalies cause contraction, and dilute acids relaxation, of the involuntary muscular fibres of the blood-vessels. Our observations show that this is also the case with voluntary muscular fibre, but, in addition, we note that acids beyond a certain strength cause a permanent contraction.

We tested the action of dilute acids and alkalies on muscle in two ways:—first by applying them directly to the muscle, and secondly by causing them to circulate artificially through the vessels supplying it. As water alone has a destructive action on muscular fibre, the acid and alkali was in all cases added to a 0·75 per cent. solution of sodium chloride.

The muscle-chamber designed by one of us (CASH), which was used in these and many other experiments, consists of a glass cylinder 3 centims. broad, 7 centims. long, and with

* This part of the paper was received June 15, 1881, but publication was deferred.

a capacity of about 40 cub. centims.   Tubes (a)* for the ingress and egress of the fluids are let into the sides of the cylinder, two above and one below.   The upper end of the cylinder is fitted accurately with a stopper made of cork and vulcanite.   The vulcanite lid (b) and the cork have an opening in the centre, which can be completely closed by means of a brass sliding clamp (c), which is moved by a screw (d) provided with a milled head.   This slide-clamp holds securely the femur, if the gastrocnemius of the Frog be used ; the ilium, if the triceps.   A binding-screw (e) is attached to the brass arm of the clamp, and this receives one of the wires of the secondary coil for direct stimulation.   The second connexion with the muscle is effected by means of a long and very fine coiled wire (f), which is in contact above with another binding-screw situated on the vulcanite cap, and below with a trout hook (g) bent into an S shape, on to which the wire is whipped.   The lower end of the S is connected with the thread or gut which passes through the lower end of the cylinder to the lever.   A second pair of binding-screws on the vulcanite lid are connected with platinum electrodes supported on a vulcanite back (h) which projects into the cylinder.   These are intended for indirect stimulation of the muscle.   Finally, the stopper carries a groove round the central opening, into which a metal cap (i) fits ; application of this cap, when the groove has been filled with a drop of oil, renders the upper opening practically air-tight.   The stopper is of course removed when a preparation for examination is placed in the chamber.   The lower end of the cylinder is permanently closed by a stopper of wood or vulcanite, which is cemented into position.   It contains two openings : the first, that of a small tube (k), through which a few drops of oil may be introduced when it is desired to make the chamber absolutely air-tight, as in experiments on the action of gases upon muscle ; the second serves for the transmission of the thread or strand of gut which connects the lever and the tendon of the muscle.   It is made from a piece of thick-walled glass tubing (l) of 1 centim. in length, drawn out with an hour-glass contraction in the middle.   The calibre at the constriction is such that a strand of very fine silk, or the best drawn trout gut just passes through it, and no more.   When the cylinder is filled with liquid the inner surface of this capillary tube becomes moistened, and it is found, whilst all friction is obviated, the escape of fluid may be reduced to such an extent that twenty or thirty drops only may flow out in the twenty-four hours.   We have repeatedly used the chamber in experiments extending over twelve hours, and found it practically full at the end of the experiment.

One of the upper openings in the wall of the cylinder is connected, by means of a T-tube, with two or more funnels, which contain : (1) the poison or poisons in solution to be tested ; (2) normal salt solution for washing out the cylinder.   The tubes connecting these with the cylinder are controlled by clamps.   In order to avoid escape of current, the fluid in the cylinder is run off before stimulation is applied. The nerve can, however, be stimulated whilst the muscle remains in the solution.

* The letters apply to both diagrams A and B, Plate 10.

Further, by regulating the height of the fluid the nerve can be exposed to the action of the solution, or kept free from it. The chamber is enclosed by a belt (*m*) connected with a rod, which fits into a nut sliding up and down on a steel upright. The lever is connected with the muscle in the usual manner, and its axis moves, together with the chamber, upon the rod of which it is clamped. By certain modifications this chamber is heated or cooled, so that the effects of variation of temperature upon the poisoned muscle may be easily studied. It is also possible to test the effect produced not only by hot and cold air, but by solutions gradually heated or cooled to any desired extent.

As already mentioned, the apparatus serves the purpose of testing the effect of gases and vapours on muscles very satisfactorily.

This mode of application was chosen on account of the obstacles to the circulation of alkalies in the muscle, and also because Von AXREP[*] asserts (1) that the action of a solution thus locally applied is the same as when the solution has been made to circulate through the tissues. GASKELL has privately communicated to us the same result, and numerous experiments of our own have confirmed these statements.

Von AXREP, in investigating the action of potassium upon muscle, found that it caused, either when applied locally or through the circulation, a decided shortening of the muscle, which in a few minutes reached its maximum. This shortening is independent of the action of the spinal cord, for it occurs whether the muscle remains in connexion with the cord, or whether the nerves be cut. The shortening has no relationship to the irritability of the muscle. The irritability of a muscle through which a 1 per cent. solution of potash is circulated for fifteen to twenty minutes is quite abolished, while the shortening persists; occasionally a slight elongation is seen, in place of a shortening. On the other hand, he found that sodium has not this effect on muscle.

Effects of Acid and Alkali applied externally to Muscles at rest.

Dilute solutions of potash and soda, containing from one part in 4,000 to one part in 8,000, cause shortening of the muscle. The contraction produced by soda was slightly greater in our experiments than that caused by potash, the solutions applied being of equal strength, and for an equal time.

Lactic acid, in very dilute solution of 1 to 8,000 or more, seems to tend to elongate muscle which is loaded with a slight weight.

A solution of chloride of sodium alone, however, also causes relaxation of the muscle, and the continuous application of a slight weight has a similar effect.

Less dilute solutions of lactic acid, 1 in 4,000 or stronger, causes passive shortening of the muscle, and this is occasionally accompanied with fibrillary twitchings. Dilute solutions of lactic acid cause relaxation of the muscle which has been shortened by potash or soda.

There is a fairly balanced antagonism between lactic acid 1 to 8,000, and soda

* PFLÜGER's Archiv., vol. xxi., p. 226.

1 to 3,000. Solutions of from 1 to 10,000 to 1 to 12,000 have both a slight power of counteracting the power of soda, and of lengthening the muscle; but 1 to 8,000 is the weakest dilution which is reliable for this purpose when applied externally. Normal salt solution has a distinct power of removing the shortening produced by soda, but its action is much more limited, and less complete than that of lactic acid.

External application of dilute acids and alkalies to contracting muscle (Plate 9, figs. 25, 26, 27). Soda and potash in solutions up to 1 in 8,000, or 1 in 10,000, cause a tonic shortening of the muscle, and may, at first, increase the height of its active contraction.

Lactic acid in dilute solutions of 1 in 10,000, or weaker, may cause elongation to a muscle which has already soaked for some time in a salt solution. A solution of 1 in 10,000 may cause at first a slight increase in the excitability and increased height of contraction, but this soon disappears. In dilutions between 1 in 8,000 and 1 in 2,000 it causes eventually shortening of the muscle, with occasional fibrillation and rapid diminution of the extent of active contraction. At the same time that the contractile power is diminishing, the muscle exhibits increasing viscosity. This is shown by a slight elevation of the basal line when the stimuli succeed each other with sufficient frequency.

The permanent shortening caused by the application of an alkali is usually diminished by the subsequent application of lactic acid. After the diminution has occurred active contraction becomes feebler.

Plate 9, fig. 25, shows the result of admitting soda solution 1 in 2,000 to the chamber containing a muscle which is being periodically stimulated through its nerve. (The solution almost entirely covers the muscle, but the nerve lying on the electrodes is above its level.) Plate 9, figs. 26 and 27, show the action of 1 to 4,000 and 1 to 5,000 soda solutions on the acting curarised muscle. Here stimulation was of course direct, and the probable escape of current is therefore to be borne in mind. In both cases the subsequent action of lactic acid is shown, viz., a reduction of the basal line, and ultimately a fall in the altitude of the contraction.

Action of Acids and Alkalies when circulated through the Muscle.

The method employed was to pith and curarise a frog. A canula was then inserted into the aorta and connected with a branching tube, through which acid, alkaline, or salt solution could be supplied from a series of funnels. By elevating or depressing the funnels the pressure by which the circulation was carried on could be increased or diminished. Excepting when otherwise stated it was always effected at as low a pressure as possible. The condition of the muscle was registered by means of MAREY'S myograph. The triceps was found to be the most convenient muscle for this series of experiments on account of its great vascularity.

Moderately dilute solutions, both of acids and alkalies 1 to 4,000, after circulating for some time, caused the muscle to shorten. Galvanic stimulation to the muscle increases this effect, both of these solutions and also of weaker ones. It frequently happens that a muscle which exhibits little or no shortening before stimulation, becomes progressively shortened after a number of stimuli have been applied, until the basal line of the curve it describes is far above the normal.

The pressure which is sufficient for the circulation of an acid solution, as a rule, quickly becomes insufficient to maintain the free circulation of an alkaline solution. This is to be expected from the fact that an alkali causes contraction of the involuntary muscular fibres of the vessels, and is in unison with GASKELL's observation.

The first effect of an alkaline solution, as a rule, is to increase the contractility of the muscle on stimulation; the same stimulus producing a greater contraction than it would in the muscle without such circulation. A gradual shortening of the muscle, independently of any active contraction, is produced by the alkaline solution: this is shown by the rise of the basal line in the curve. After the circulation has been maintained for some time, both the contractile power and the irritability of the muscle decrease; the height of the contraction occurring on stimulation not being so great, and a stronger stimulus being required.

Plate 9, fig. 28, a, b, c, is introduced to show the fibrillation and temporary shortening which may occur upon the first stimulations of a muscle through which lactic acid has been some time circulated.

Plate 9, fig. 29, a, b, c, shows that the elevation of the basal line, caused by the circulation of soda (1-20,000), is to a large extent reduced by the subsequent circulation of lactic acid 1-10,000. The altitude of the contraction is likewise reduced.

Lactic acid, when circulated through the muscle, frequently causes fibrillation, and at first shortening of the muscle after fibrillation: there may, however, not be any shortening.

Usually the height of the contractions diminishes rapidly on repeated stimulation; sometimes, though quite exceptionally, the irritability of the muscle is increased at first, and the contractions resulting from stimulation may be at first more extensive than those of the normal muscle.

Œdema of the muscle is occasionally observed as a consequence of the circulation of acid through the vessels; this is unusual after the circulation of alkalies. The impaired contractile power eventually produced by the circulation of either alkali or acid through a muscle may be restored to a greater or less extent by the circulation of a fluid having an opposite reaction. The completeness of the restoration depends upon various circumstances, amongst which we may mention the œdematous condition of the muscle, which we have already noticed as occurring from the circulation of acids.

Our experiments on the muscles of the Frog have thus shown a very marked antagonistic power between acids and alkalies, or perhaps to speak more definitely, between solutions of potash or soda and lactic acid. It seemed advisable to make some experiments on the muscles of warm-blooded animals, in order to discover whether the same antagonism was to be found in them: for this purpose we chose the gastrocnemius of the Cat. The solution to be investigated was warmed to 40° C, and then passed through the limb by means of a canula inserted into the femoral artery. The muscle was stimulated from the sciatic nerve, the leg being previously fixed by a clamp. The muscle was extended by a weight of 40 grammes attached by a cord working over a pulley ; this was allowed to remain constantly attached in some experiments to ascertain alterations in the length of the muscle due to the fluids circulated. In several cases it was applied for two minutes before each tracing.

Plate 9, fig. 30, shows the effect of acids and alkalies.

(a.) The lever recorded (multiplies 4 times) contractions of 12·5 millims., an opening and closing shock every 4".

(b.) After alkali 1-20,000 had circulated 10ᵐ the basal line showed a shortening of 8 millims. The active contraction was 13 millims. Ten minutes after this tracing had been taken the flow, which had previously been free from the femoral vein, became very slow, and remained so under a considerable increase of pressure.

(c.) Lactic acid 1-10,000 restored the circulation and reduced the contraction. The active contraction of the value of 12 millims.

(d) Alkali circulated 20ᵐ has raised the basal line 10·5 millims., but shows an active contraction of less than 10 millims.

(e.) After 60ᵐ circulation the basal line is still 10·5 above the normal, but the active contraction has increased to 11·5 millims.

There is here, then, a great similarity of action in the case of acid and alkali circulated through the vessels of cold and warm-blooded animals.

General Results of Experiments on the Action of Acid and Alkali on Muscle.

The experiments just described show that dilute alkalies, potash, and soda cause shortening of muscle, which is antagonised by dilute solution of lactic acid. Since the preceding section of this paper was sent in to the Royal Society we have made some further observations on this subject, and from an examination of the curves it will be seen that, by the alternate application of alkali and acid, a muscle may be made to describe on a slowly revolving cylinder a curve very nearly resembling that described on a rapidly revolving cylinder by a normal muscle when stimulated. Other tracings show that this curve may be modified very nearly at will by altering the proportions and duration of the alkali and acid. Curves may be thus described which resemble those drawn by muscles stimulated after they have been poisoned by barium, rubidium, and other substances of the groups we have examined. In these curves we see produced by varying the application of the opposing solutions the same prolonged contraction, the tendency to an exaggerated secondary hump, and increased contracture.

We cannot at present draw from this a definite conclusion, but it is suggestive of the question—Does the normal contraction of muscle and its subsequent relaxation depend upon such alterations in its saline constituents as to make them play at one time the part of an alkali, and at the other the part of an acid ?

Plate 9, figs. 31 and 32, show the relative effects of solutions of 1 to 3,000 caustic soda solution (Plate 9, fig. 31) and caustic potash solution (Plate 9, fig. 32) upon resting muscle. The tracings were taken upon a slowly revolving cylinder. Each centimeter of the tracing represents 5". The lever, which multiplies fourteen times, exercises a constant traction of 10 grms. on the muscle. Fresh solution was added where stars are placed in the course of the curve. It will be seen that the shortening effect produced by caustic soda in 50", during which the solution was renewed every 10", is slightly greater than is the case with the companion muscle treated with caustic potash of the same strength. The curves, however, show a very close similarity throughout. The commencing relaxation caused by the substitution of 1 to 1,000 lactic acid is seen in each case.

The very gradual shortening of the muscle upon the first application of potash and soda is, to some extent, due to the fact that the muscles had been previously curarised. When curara has not been previously employed the first application of dilute solutions causes a more rapid primary contraction, though the total effect of the application may not be greater, if as great as in the curarised muscle. Plate 9, fig. 33, gives the effect of a stronger solution of soda, i.e., 1 to 2,500, and the subsequent relaxation it undergoes upon the application of 1 to 500 lactic acid.

Plate 9, figs. 34 and 35, give the action of soda 1 to 4,000, and potash 1 to 6,000, with partial relaxation consequent to lactic acid. That lactic acid itself causes shortening, if of a certain strength, is shown in Plate 9, fig. 36, when 1 to 1,000 solution of the acid causes in 25" a shortening of 4 millims. in the curve, or of ·3 millim. in the muscle.

The application of potash reduces this shortening to some extent, and then, its own action being no longer balanced, causes the muscle to contract rapidly. The converse of this is seen in Plate 9, fig. 37, when the alkali is first applied, and the acid 1 to 500 causes a relaxation, and then a shortening of its own. To cause a complete relaxation a higher dilution is necessary.

Plate 9, figs. 38 and 39, give tracings of passive shortening or lengthening with an active contraction (maximal stimulation) taken at intervals superimposed.

Plate 10, fig. 40, a, b, c, illustrates the change of form the normal muscle curve undergoes when treated with an alkali local application. The first "hump" of the active contraction is increased in altitude; the second "hump" or elevation after the notch is reduced. Owing to this reduction the curve is shortened. A passive shortening of the muscle is seen at c, and is, in point of fact, less than is usually produced by solutions of these strengths.

The effect of lactic acid applied in the same manner is shown in the series a, b, c, Plate 10, fig. 41. Here also the second portion of the curve is reduced, and the relaxation becomes much more rapid. After 60" in lactic acid 1 to 2,500, a slight contraction of 1·5 millim. is observable.

Plate 10, fig. 42, a, b, c, d, e, gives the action of potash on the normal muscle, to a large extent counteracted by lactic acid, and the subsequent passive shortening of the muscle under the non-balanced action of a strong solution (1 to 500) of the acid.

ON THE RELATIVE ACTION OF ALKALIES AND ALKALINE EARTHS ON MUSCLE.

We cannot enter here into a full consideration of the antagonism which certain members of these groups show with regard to the action of other members, but we may briefly state a few of the most striking facts. Thus potassium shortens the lengthened curves of veratria, barium (Plate 10, fig. 43), calcium, strontium, of large doses of sodium and of lithium (Plate 10, fig. 44), and reduces the contracture which they have caused. Sodium, which we have shown in large doses to cause a lengthened curve with increased contraction, adds to the length of calcium and strontium when applied in strong solutions. Barium, when it has produced its lengthened veratria-like curve, is, however, counteracted by almost all the substances which tend to produce a shorter curve. Thus calcium and potassium both of them lessen its altitude, and abolish its contracture. A remarkable antagonism, however, is that existing between rubidium and barium. The veratria-like curve which the former has been shown to cause when in strong solution is completely reduced by the application of a solution of barium, of such a strength as would, if applied by itself in the first instance, have caused a similar, though more extensively varied, curve. It is to be noted that in this antagonism, as in many others, the muscle yields a reaction closely similar to the normal before it develops the characteristic curve which is associated with the substance used to antagonise.

With two substances of closely-allied action we sometimes find, as in the case of calcium and strontium, an addition of effect (Plate 10, fig. 45) without any reduction having taken place. It would appear that in some cases we get the two substances which have a similar action, at one time aiding one another, in other cases neutralising one another. It is hard to say what the cause of this curious result is, and any explanation of it must be at present entirely hypothetical. At present our data are too limited to allow us to formulate any general rule regarding antagonism. We may, however, mention some antagonisms which are at any rate curious.

(1) Calcium reduces the barium curve to the normal, or thereabouts, before it causes its own peculiar form of curve.

(2) Rubidium in strong solutions has the same effect as barium in causing a veratria-like curve.

(3) Sodium usually produces with lime, not a shortening of the curve, but an increase of the after-action (contracture) which is often seen in the lithium muscle.

(4) Potash lengthens the curves of didymium and lanthanum.

(5) Lithium increases calcium effect, and calcium increases lithium effect.

(6) Potassium opposes strontium.

(7) We have drawn attention to the antagonism of barium to rubidium (when the

latter develops in strong solution a veratria-like curve), and also that potassium is antagonistic to barium.

(8) Sodium, in strong solutions, may reduce the lithium contraction before the death of the muscle occurs.

Although we have at present considered the action of ammonia, compound ammonias, alkalies, and alkaline earths, on voluntary muscle only, we have made a number of experiments which seem to show that their action on involuntary muscular fibre is very similar, e.g., barium causes a very great prolongation of systole in the Frog's heart, just as it prolongs the contraction of voluntary muscle. These results we intend to investigate more fully, and hope to publish them hereafter.

All attempts to establish a relationship between atomic weight and physiological action have hitherto failed. It may be that this failure has resulted from the lethal activity on the organism, as a whole, having been taken into consideration, whereas different substances may cause death by acting on different structures. We think that by the method here pursued of investigating their relationship to one or two structures only, and by a careful comparison of their actions, some definite connection may yet be established, and we hope that the results which have been recorded may serve as a contribution towards this end.

Perhaps they may also serve to throw some light on the curious subject of the different reactions of different organisms to the same drug, but this also we purpose to follow up in a further research.

We desire to acknowledge most gratefully the great kindness of Professor RANVIER, who placed his laboratory at our disposal, and afforded us every facility for carrying out there the experiments on warm-blooded animals, and also on unpithed Frogs, which are rendered so difficult in this country by the present state of the law.

EXPLANATION OF FIGURES.

PLATE 8.

The figures represent the curves obtained by registering the contraction of the gastrocnemius of the Frog (*Rana Temporaria*) on a revolving cylinder.

Fig. 1. Frog poisoned by 1 drop 10 per cent. solution of dimethyl-ammonium chloride.
　　a. Ligatured leg.　5ˢ tetanus, direct stimulation of gastrocnemius.
　　b. Ditto.　Indirect stimulation.
　　c. Poisoned leg.　Direct stimulation.
　　d. Ditto.　Indirect stimulation.

Fig. 2. Frog poisoned by tetramethyl-ammonium iodide.

    a. Ligatured leg. Ten stimulations* (direct) of gastrocnemius, one stimu-
            lation every 1·5°.

    b. Poisoned leg.   Ditto.

Fig. 3. Frog poisoned by large dose (·2 grm.) amyl-ammonium iodide.

    a. Ligatured leg. Ten stimulations (direct) of gastrocnemius, one stimula-
            tion every 1·5°.

    b. Poisoned leg.   Ditto.

    c. Poisoned leg.   Single curve, indirect stimulation.

Fig. 4. Frog poisoned by trimethyl-ammonium iodide.

    a. Ligatured leg.   Ten stimulations (direct) of gastrocnemius ; curves of
            direct and indirect stimulation are equal ; one stimulation
            every 1·5°.

    b. Poisoned leg.   Ditto.   Direct stimulation.

    c. Poisoned leg.   Ditto.   Indirect stimulation.

Fig. 5. Frogs poisoned by tetraethyl-ammonium iodide.

    a. Ligatured leg.   Single stimulation of gastrocnemius (direct).

    b. Poisoned leg.   Ditto.   The nerve is no longer irritable.

    c. Case of profound poisoning.   Direct stimulation of gastrocnemius.

Fig. 6. Frog poisoned by dimethyl-ammonium sulphate (·25 grm.).

    a. Ligatured leg.   Direct and indirect stimulation.

    b. Poisoned leg.   Direct stimulation.   Nerve no longer irritable.

Fig. 7. Frog slightly poisoned by trimethyl-ammonium sulphate (·1 grm.).

    a. Ligatured leg.   Tetanus 5°, direct stimulation.

    b. Ligatured leg.   Ditto, indirect stimulation.

    c. Poisoned leg.   Tetanus 5°, direct stimulation.

    d. Poisoned leg.   Ditto, indirect stimulation.

Fig. 8. *a.* Normal gastrocnemius.   Direct stimulation.

    b. Ditto.   After 5ᵐ in 1 per cent. chloride of rubidium solution.

    c. Ditto.   After 15ᵐ in ·75 per cent. chloride of calcium solution.

Fig. 9. *a.* Normal gastrocnemius.   Direct stimulation.

    b. Ditto.   After 20ᵐ in 1–1000 chloride of ammonium solution.

Fig. 10. *a.* Normal gastrocnemius.   Direct stimulation.

    b. Ditto.   After 30ᵐ in 2 per cent. solution chloride of sodium.

    c. Ditto.   After 45ᵐ in ditto.

Fig. 11. Frog poisoned by ·02 grm. chloride of cæsium.

    a. Ligatured leg.   Direct stimulation.

    b. Poisoned leg.   Ditto.

Fig. 12. *a.* Normal gastrocnemius.   Direct stimulation.

    b. Ditto.   After 15ᵐ in 1 per cent. chloride of ammonium.

          * All single stimulations are by an opening maximal induction shock.

Fig. 13. *a.* Normal gastrocnemius. Direct stimulation.
  b. Ditto. After 30ᵐ in ·33 per cent. chloride of lithium.
Fig. 14. *a.* Normal gastrocnemius. Direct stimulation.
  b. Ditto. After 30ᵐ in ·1 per cent. chloride of potassium.
  c. Ditto. After 30ᵐ in ·15 per cent. ditto.
Fig. 15. *a.* Normal gastrocnemius. Direct stimulation.
  b. Ditto. After 30ᵐ in ·25 per cent. chloride of barium.
  c. Ditto. After 45ᵐ in ditto.
  d. Ditto. After 15ᵐ in ·25 per cent. chloride of potassium.
Fig. 16. Frog poisoned by chloride of erbium (slow action of drug).
  a. Ligatured leg. Direct stimulation.
  b. Poisoned leg. Direct and indirect stimulation give equal contractions.
Fig. 17. Frog poisoned by chloride of lanthanum.
  a. Ligatured leg. Indirect stimulation.
  b. Poisoned leg. Indirect stimulation.
Fig. 18. Frog poisoned by chloride of yttrium (slow action of drug).
  a. Ligatured leg. Indirect stimulation.
  b. Poisoned leg. Ditto.
Fig. 19. Frog poisoned by ·35 grm. calcium chloride.
  a. Ligatured leg. Indirect and Direct stimulation give equal contractions.
  b. Poisoned leg. Indirect stimulation.
  c. Ditto. Direct stimulation.

PLATE 9.

Fig. 20. *a.* Normal gastrocnemius. Direct stimulation.
  b. Ditto. After 20ᵐ in 1 per cent. chloride of didymium.
Fig. 21. *a.* Normal gastrocnemius. Direct stimulation.
  b. Ditto. After 30ᵐ in ·2 per cent. chloride of strontium.
  c. Ditto. After 15ᵐ in ·5 per cent. ditto.
Fig. 22. *a.* Normal gastrocnemius. Direct stimulation.
  b. Ditto. After 20ᵐ in 1 per cent. chloride of beryllium.
Fig. 23. Frog poisoned by beryllium chloride (·02 grm.).
  a. Ligatured leg. Tetanus of gastrocnemius, direct stimulation.
  b. Poisoned leg. Ditto. Secondary coil at 2 c.m. Indirect stimulation of the poisoned muscle did not yield any contraction.
Fig. 24. Action of heat and cold on the barium curve.
  a. Normal gastrocnemius. Direct stimulation, at room temperature 13° C.
  b. Ditto. After 15ᵐ in ·25 per cent. chloride of barium solution. Temperature 13° C.

      c. Ditto. Application of barium solution continued. Kept for 15^m, cooled
          to 8°·5 C.

      d. Ditto. Heat to 18° C. Reappearance of veratria-like curve.

      e. Ditto. Heat to 20° C.

      f. Ditto. Heat to 30° C. The veratria-like curve disappears.

      g. Ditto. Cool to 14° C. There is no return to the veratria-like curve. A
          simple prolonged contraction persists.

Fig. 25. Action of soda on contracting muscle. Solution of 1–2000 admitted at ×.
      Stimulation every 2".

Fig. 26. *a.* Action of soda, 1–4000, on contracting curarised muscle. Solution
      admitted at ×.

      b. Same muscle after exposure to lactic acid, 1–4000, for 40^m. Stimulation
      every 2", direct.

Fig. 27. *a.* Action of soda, 1–5000, on contracting curarised muscle. Solution
      admitted at ×.

      b. Lactic acid, 1–5000, has acted 1^m on muscle.

      c. Ditto, has acted 5^m on muscle. Stimulation every 2", direct.

Fig. 28. *a.* Normal gastrocnemius. One opening and one closing stimulation every 4".

      b. After 60^m circulation of lactic acid through aorta, 1–8000, stimulation
      causes fibrillation and shortening of muscle.

      c. After 30^m circulation of soda, 1–6000, the strength of contraction, which
      had been diminished under acid, is restored; fibrillation has ceased.

Fig. 29. *a.* Normal gastrocnemius. One opening and one closing stimulation every 4".

      b. Taken after circulation for 10^m of 1–20,000 alkaline solution.

      c. Taken after circulation for 30^m of 1–10,000 acid solution.

Fig. 30. Tracing from gastrocnemius of Cat. One opening and one closing stimulation
      every 4". The solution, heated to 38° C., was circulated under pressure
      through the femoral artery, and allowed to escape by the femoral vein.
      The rest of the limb, with the exception of the sciatic nerve, which was
      exposed for stimulation, was ligatured. A weight of 40 grms. was
      applied 2^m before each tracing was taken. Abscissæ constant.

      a. Normal contractions.

      b. Alkali, 1–20,000, has circulated 10^m.

      c. Acid, 1–10,000, has circulated 60^m.

      d. Alkali, as before, 20^m.

      e. Ditto, 60^m. Flow from venous canula very slow and weak.

Fig. 31. Action of alkali and acid upon resting muscle (curarised). At the first five
      points indicated by ×, soda solution, 1–3000, is supplied to muscle in
      cylinder. At the last six points indicated by a ×, lactic acid, 1–1000,
      is supplied. The action of the soda was for 47·5^m; that of the acid for
      42^m. Change of alkali to acid, or *vice versâ*, in all cases shown by
      double-headed arrow.

Fig. 32. Action of caustic potash, 1–3000, for 43ᵐ, succeeded by action of lactic acid, 1–1000 for 42ᵐ.

Fig. 33. Caustic soda, 1–2500, once renewed in 25ᵐ, succeeded by action of lactic acid, 1–500, once renewed in 25ᵐ.

Fig. 34. Curarised gastrocnemius. Caustic soda, 1–4000, twice renewed in 33ᵐ, succeeded by action of lactic acid, 1–1500, once renewed in 25ᵐ.

Fig. 35. Curarised gastrocnemius. Caustic potash, 1–6000, thrice renewed in 46ᵐ, succeeded by lactic acid, 1–1500, twice renewed in 28ᵐ.

Fig. 36. Curarised gastrocnemius. Lactic acid, 1–1000, four times renewed in 37ᵐ, succeeded by caustic potash, 1–2500, once renewed in 34ᵐ.

Fig. 37. Action of caustic potash, 1–2500, twice renewed for 13ᵐ, succeeded by action of lactic acid (1–500) for 18ᵐ, and this by action of caustic potash for 17·5ᵐ.

Fig. 38. Action of caustic potash, 1–4000, for 20ᵐ, succeeded by lactic acid, 1–1000, 48ᵐ. The muscle is subjected to maximal stimulation before the change of each solution.
   1. Contraction of normal muscle.
   2, 3. Contractions of alkali muscle.
   4, 5, 6, and 7. Contractions of acid muscle.

Fig. 39. Action of potash, 1–1500, for 18ᵐ, succeeded by lactic acid, 1–500, for 24ᵐ.
   1. Contraction of normal muscle.
   2, 3, 4. Contractions of alkali muscle.
   5, 6, 7. Contractions of acid muscle.

PLATE 10.

Fig. 40. a. Curve of normal gastrocnemius. Direct stimulation.
   b. Ditto. After 10ᵐ in soda solution, 1–3000.
   c. Ditto. After 20ᵐ in ditto.

Fig. 41. a. Curve of normal gastrocnemius. Direct stimulation.
   b. Ditto. After 15ᵐ in lactic acid solution, 1–2500.
   c. Ditto. After 30ᵐ in ditto.

Fig. 42. a. Curve of normal gastrocnemius. Direct stimulation.
   b. Ditto. After 15ᵐ in potash solution, 1–4000.
   c. Ditto. After 15ᵐ in lactic acid solution, 1–500.
   d. Ditto. After 30ᵐ in ditto.
   e. Ditto. After 45ᵐ in ditto.

Fig. 43. *a*. Curve of gastrocnemius which has been 20^m in barium chloride solution, 1–600.

   b. Ditto.   After 15^m in chloride of potash solution, 1–600.

   c. Ditto.   After 30^m in ditto.

Fig. 44. *a*. Curve of gastrocnemius which has been 80^m in chloride of lithium solution, 1–300.

   b. Ditto.   After 15^m in chloride of sodium solution, ·75 per cent.

   c. Ditto.   After 30^m in ditto.

   d. Ditto.   After 15^m in chloride of potassium solution, 1–800.

Fig. 45. *a*. Curve of normal gastrocnemius.   Direct stimulation.

   b. Ditto.   After 30^m in chloride of strontium solution, 1–150.

   c. Ditto.   After 15^m in chloride of calcium solution, 1–150.

Diagrams of muscle chamber, A and B.

a, a. Influx and efflux tubes for solutions.

b. Vulcanite lid cemented into cork which closes the upper end of the chamber.

c. Sliding clamp which fixes the femur moved by milled-headed screw (*d*).

e. Clamp for carrying wire for direct stimulation. The second connexion is made through the coiled wire (*f*), terminating in a hook (*g*) which passes through the tendon of the muscle.

h. Electrodes for stimulation of the nerve.

i. Metal cap closing central opening in stopper.

k. Accessory escape or oil tube.

l. Tube with hour-glass contraction, through which thread connecting tendon and lever works.

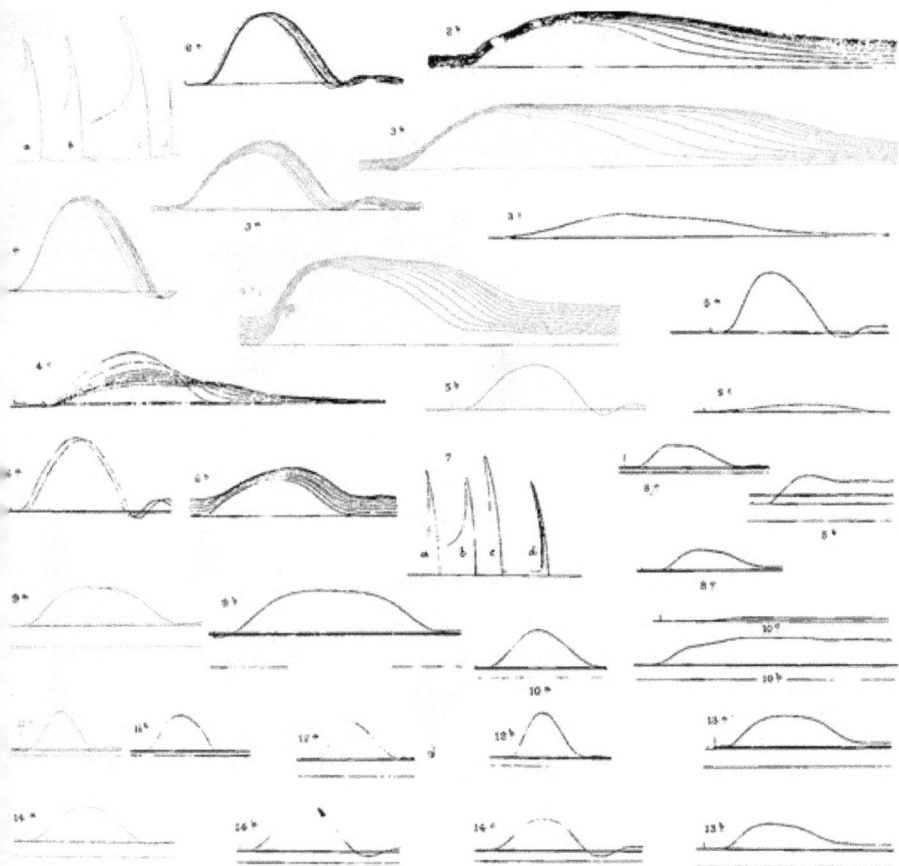

Brunton & Cash

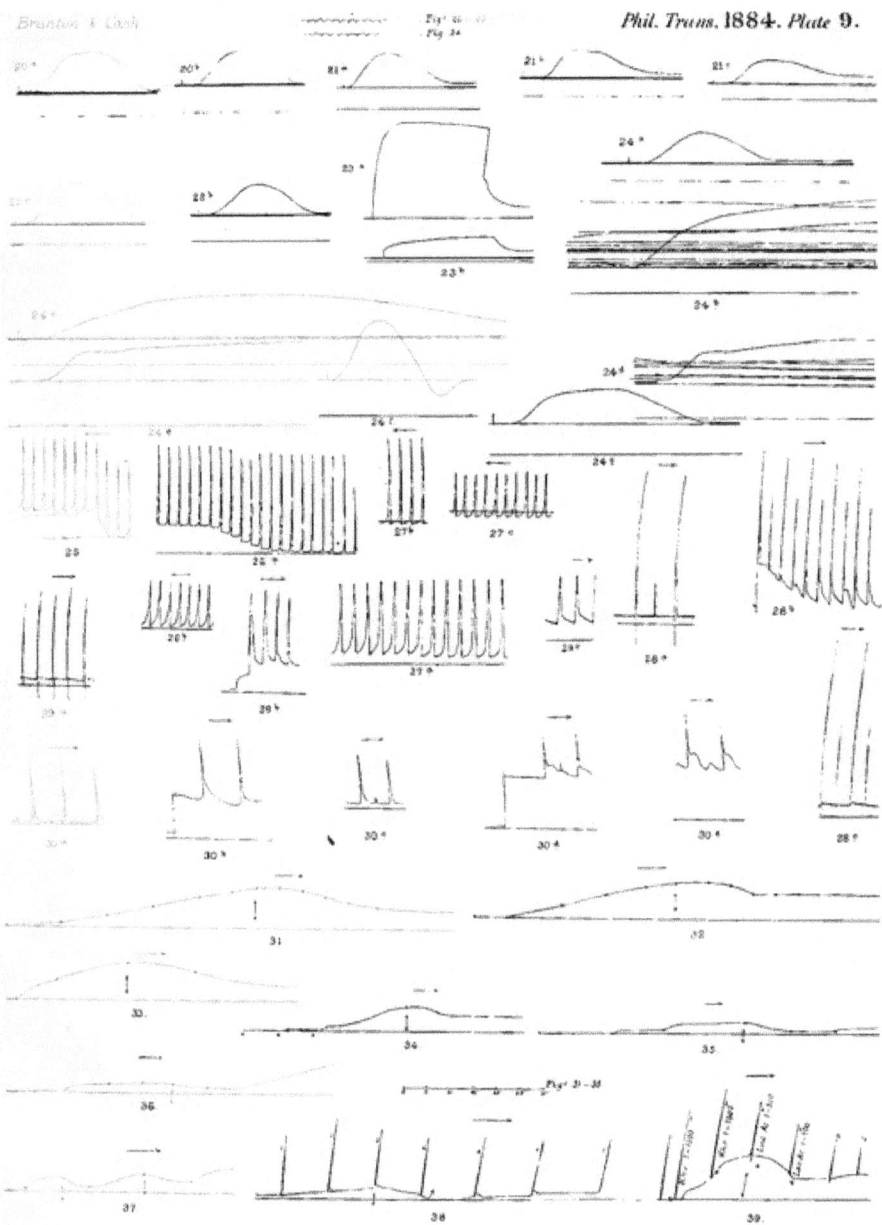

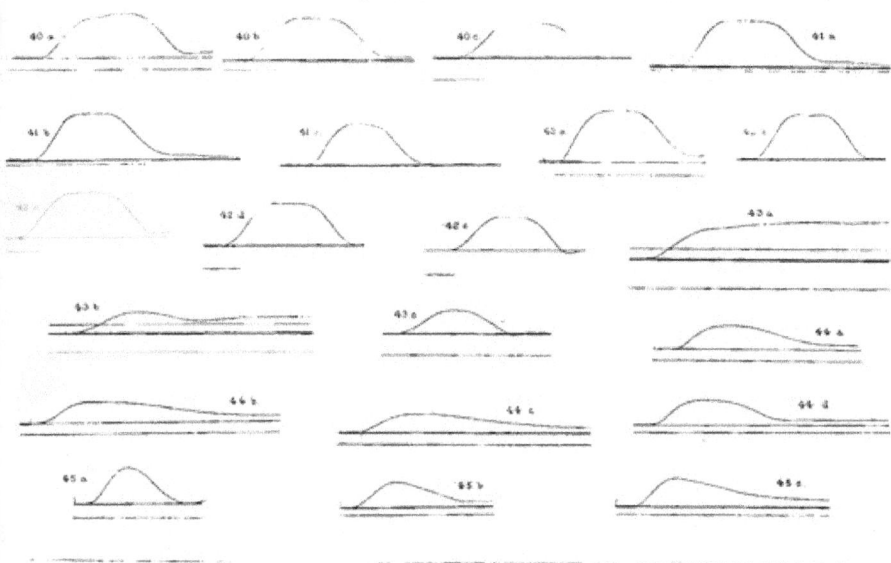

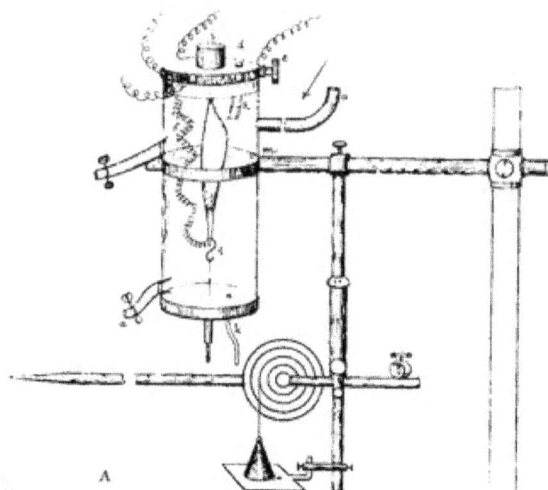

Drawing of direct Muscle Chamber
with Lever in connection
For explanation of Figures see text

A

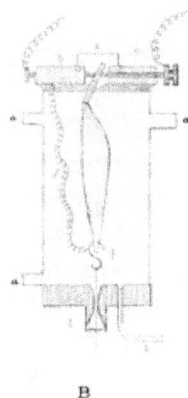

Diagramatic Section of closed
Muscle Chamber (longitudinal)
For explanation of Figures see text

B

www.ingramcontent.com/pod-product-compliance
Lightning Source LLC
Chambersburg PA
CBHW022038080426
42733CB00007B/877